Empowerment

For Leonie

EMPOWERMENT

The Politics of Alternative Development

John Friedmann

BLACKWELL
Cambridge MA & Oxford UK

First published 1992

Blackwell Publishers
Three Cambridge Center
Cambridge, Massachusetts 02142
USA

108 Cowley Road
Oxford OX4 1JF
UK

Library of Congress Cataloging-in-Publication Data

Friedmann, John.
    Empowerment: the politics of alternative development/John
Friedmann.
        p.   cm.
    Includes bibliographical references and index.
    ISBN 1–55786–299–0. — ISBN 1–55786–300–8 (pbk.)
    1. Community development.   2. Economic development—Social
aspects.   I. Title.
    HN49. C6F75   1992
    307. 1′4—dc20   91–23631
                                                                          CIP

British Library Cataloguing in Publication Data

A CIP catalogue record for this book is available from the British Library.

Typeset in 10 on 12 pt Sabon
by Graphicraft Typesetters Ltd., Hong Kong
Printed in the USA

This book is printed on acid-free paper

Contents

Preface

This book is about failure, and about hope. The failure is that of main-stream models of economic development: their inability to address the massive problems of world poverty and environmental sustainability. The hope is the emerging practice of an alternative development with its claims to inclusive democracy, appropriate economic growth, gender equality, and intergenerational equity. *"Failure"* and *"hope"* are not usually part of the vocabulary of analytical social science. They are moral and normative words. Here they express a bias for people as active subjects of their own history. People's empowerment – their collective self-empowerment – lies at the heart of the practice of an alternative development.

My intention is to restate the aims and practices of an alternative development in light of the experience gathered over more than 20 years. How are we to assess this experience? What have we learned? The task is made more difficult by the absence of texts that attempt to summarize an alternative doctrine. There is a large, mostly ephemeral literature on nongovernmental organizations that have become identified with an alternative development. This literature, discussed in chapter 7, is largely exhortatory. There are also numerous writings on basic needs, sustainability, and women in development, some of which are discussed in later chapters. But collectively the writings on alternative development tend to be fragmented and dispersed. With the possible exception of Ross and Usher's *From the Roots Up: Economic Development as if Community Mattered* (1986), there are no systematic statements of an alternative development doctrine that one could critically examine. And Ross and Usher address problems of local development in a North American context. They are not specifically interested in Third World poverty.

I have therefore written this book to provide a theoretically coherent as well as morally informed framework that can serve as a point of departure for historical practice. The time for such an undertaking is auspicious. History is racing forward. A truly global economy is in the making. New technologies are coming on stream, erasing time and distance. Old agrarian economies in Southeast Asia are industrializing, while old industrial regions in Eastern Europe and elsewhere are scrambling to modernize and link up with the new globalism. With the imminent breakup of the Union of Soviet Socialist Republics, an era of empires is drawing to a close. Central planning and bureaucratic direction are yielding to more flexible, decentralized, fine-boned structures of decision-making. Universalist epistemologies are being undermined by a "postmodern" skepticism about grand theories and supposedly immutable structures. Above all, civil society has been stirred to action. In small ways and in large, from Tiananmen Square to the Plaza de Mayo in Buenos Aires, from Soweto to Vilnius, civil society is engaged in a politics of emancipation. An alternative development is part of this politics.

What we now call "civil society" – the term itself goes back to seventeenth-century political philosophers – refers to those associations beyond the reach of the state and corporate economy which have the capacity for becoming autonomous centers for action. Strictly speaking, civil society should always be used in the plural form to acknowledge the many ways it is structured: by class, caste, religion, language, regional identity, culture, race, and gender. Yet what confronts us is not chaos but a complex, many-stranded order that arises from those values and interests which, in any particular region of the world, are shared by most of its members: a burning desire for freedom from oppression and for political community.

Alternative development, as an expression of a militant civil society, is political to the core. Its politics asserts universal human rights and the particular rights of citizens in given political communities, especially the rights of the people heretofore without voice, the disempowered poor, who constitute a majority.

The ideas for an alternative development took shape some 20 years ago over the course of successive meetings of an "invisible college" of international development specialists. Disaffected by the mainstream model with its emphasis on rapid cumulative growth, its urban bias, and the single-minded pursuit of industrialization, these specialists succeeded in formulating the outlines of an approach that they hoped would lead to a direct improvement in the conditions of the poor, especially the rural poor, and at the same time be compatible with emerging environmental concerns. The currently visible incarnation of this approach is found in the explosive proliferation of nongovernmental organizations and their activities worldwide.

The empowerment approach, which is fundamental to an alternative development, places the emphasis on autonomy in the decision-making of territorially organized communities, local self-reliance (but not autarchy), direct (participatory) democracy, and experiential social learning. Its starting point is the locality, because civil society is most readily mobilized around local issues. But local action is severely constrained by global economic forces, structures of unequal wealth, and hostile class alliances. Unless these are changed as well, alternative development can never be more than a holding action to keep the poor from even greater misery and to deter the further devastation of nature. If an alternative development looks to the mobilization of civil society at the grass roots or, as Latin Americans like to say, in communities "at the base," it must also, as a second and concurrent step, seek to transform social into political power and to engage the struggle for emancipation on a larger – national and international – terrain.

An alternative development is essentially a dialectical ideology and practice. It is what it is because mainstream doctrine exists, just as the state exists. Its aim is to replace neither the one nor the other but to transform them both dramatically to make it possible for disempowered sectors to be included in political and economic processes and have their rights as citizens and human beings acknowledged. This dialectic runs across the spectrum of alternative development concepts. One's autonomy is limited by the autonomy asserted by others; self-reliance takes place in a context of interdependence, participatory democracy at the base is engaged in the larger processes of representative governance; experience-based learning is in creative tension with theoretical knowledge. The politics of an alternative development cannot be totalized. It is a transforming politics that will itself be transformed in practice.

The argument in the following chapters begins with an account of an alternative development and its moral justification. It proceeds in chapter 2 to explain the dynamics by which most of the world's population is excluded from economic and political participation. Chapter 3 follows with a critique of the widely used model of national economic accounts and argues for a rethinking of the economic foundations of an alternative development in the "whole" economy, a model that articulates market and nonmarket relations in the production of life and livelihood through the agency of the household economy. Chapter 4 then looks more closely at the concept of poverty, contrasting the traditional bureaucratic definition of the poor by an arbitrarily selected minimum level of consumption with a proposed (dis)empowerment model. If poverty is a condition of relative disempowerment with respect to a household's access to specified bases of social power, then a key to the overcoming of mass poverty is the social and political empowerment of the poor.

The next two chapters lay out the programmatic objectives of a politics of claiming by the disempowered: the claims for inclusive democracy and appropriate economic growth (chapter 5) and the claims for gender equality and sustainability (chapter 6). The concluding chapter looks critically at the actual practices of an alternative development. It examines the growing role of nongovernmental agencies and wonders what will happen when they themselves become (as some of them are already becoming) bureaucratic organizations mediating between the state and the poor. The chapter closes with an affirmation of unmediated community-based struggles. The Epilogue reminds us that poverty, exclusion, and disempowerment exist in rich countries as well as poor and raises questions for rich countries that are suggested by experiences with alternative development in the Third World.

Historical illustrations of my argument are for the most part drawn from Latin America, the region I know best, and where most of the research for this book was carried out. Properly interpreted in the context of other cultures and historically evolved social and political systems, the Latin American experience will, I hope, turn out to be useful in the emancipatory struggles of civil society in other parts of the world.

Pacific Palisades, California
June 1991

Acknowledgments

The research underlying this book was made possible by successive grants from the Academic Senate and the Center for Latin American Studies at UCLA. The Inter-American Foundation enabled me to visit a number of Latin American countries in 1987, a trip that in many ways proved decisive for my subsequent work. The bulk of the actual writing was accomplished while I was a visiting scholar at the Institute of Urban and Regional Development at the University of California at Berkeley during the 1990 spring semester. I am especially grateful to the institute's director, Peter Hall, and his administrative staff for making my visit with them so productive.

Students and former students of mine, including Simon Fass, Francisco "Pancho" Sabatini, Mauricio Salguero, Alfonso Rivas, Bishwapriya Sanyal, and Stephanie Pincetl, contributed with sharp critiques of an early version of the present text. Most of all, I am indebted to David Seddon and an anonymous critic for the great care they took in reading through the manuscript at an intermediate stage. Their extensive critical comments and observations were most helpful in my undertaking of final revisions. Although I could not always agree with them, I found what they had to say invaluable in producing the present version.

Finally, David Van Arnem at the Institute of Urban and Regional Development at Berkeley and Debra Sulkin at UCLA worked long hours putting the text on a word processor and never tired in their efforts throughout many revisions to produce ever more perfect copy. The diagrams were drafted with great care by my daughter, Manuela.

What we desperately need today is to learn to think and act more like the fox than the hedgehog – to seize upon those experiences and struggles in which there are still glimmerings of solidarity and promise of dialogic communities in which there can be genuine mutual participation and where reciprocal wooing and persuasion can prevail. For what is characteristic of our contemporary situation is not just the playing out of powerful forces that are always beyond control, or the spread of disciplinary techniques that are always beyond our grasp, but a paradoxical situation where power creates counter-power and reveals the vulnerability of power, where the very forces that undermine and inhibit communal life also create new, and frequently unpredictable, forms of solidarity.

Richard Bernstein
Beyond Objectivism and Relativism

Development is the possibility open to all of a country's inhabitants to enjoy material and spiritual prosperity instead of being merely on the "treadmill" in the machinery of accumulation. Yes, there should be economic growth, but not at the expense of the survival of the poorest.

Editorial "Revista EL SALVADOR" (En Construcción)
August 1989

1 Alternative Development: Its Origins and Moral Justification

Origins

"What has been happening to poverty? What has been happening to unemployment? What has been happening to inequality? . . . If one or two of these central problems have been growing worse, especially if all three have, it would be strange to call the result 'development,' even if per capita income doubled." In asking these questions more than 20 years ago, Dudley Seers, director of the prestigious Institute of Development Studies at the University of Sussex, called for a major rethinking of development doctrine (Seers, 1969, p. 3).[1]

Many people today know the sixties as the decade of a "movement politics" that stirred the world from Beijing to Paris. But by the end of the decade, it seemed as if the popular tide had receded and that states had regained control, limiting the course of political practice to its customary channels. This return to normalcy was more deceptive than real, however. Movement politics, as it turned out, was here to stay. The only thing that changed was the nature of the movements themselves. The new social movements of ecology, peace, and women – although perhaps less spectacular than China's convulsive Cultural Revolution, America's Black Power movement, and the Paris student uprising of May 1968 – kept up the pressure. In a historical perspective, one can see in all of them the rise of civil society as a collective actor, working for political agendas outside the established framework of party politics. Throughout the world, social movements have helped to bring about a profound democratization of politics (Castells, 1983; Touraine, 1977, 1981).

Concurrent with these broadly based social movements, an intellectual

[1] The intellectual history of alternative development approaches has not yet been written. Development economics, with which it is closely linked, has gone its own way and is beginning to have a small literature on its own "growth and decline" (Hirschman, 1981; Meier and Seers, 1984; Arndt, 1987; Evans and Stephens, 1988, Hunt, 1989). This chapter is in no sense intended to close the gap. It deals with "origins," tracing the main lineage, but does not pretend to tell the story in all its drama and richness.

movement limited mostly to scholars and development professionals fought for an alternative approach to the development of poor countries. Its achievements were won not in the streets but at international conferences whose participants took up Dudley Seers's call to give new meaning to development and to reset the policy agenda. Landmark meetings were the Stockholm Conference on the Human Environment (1972), which led to the establishment of the United Nations Environment Program (UNEP), and the Cocoyoc (Mexico) seminar on "Patterns of Resource Use, Environment and Development Strategies," convened in October 1974 at the behest of UNEP and the United Nations Conference on Trade and Development (UNCTAD). Cocoyoc brought together experts from all parts of the world, including such lustrous names as Wassily Leontief, Vladimir Kollontai, Josef Pajestka, Ignacy Sachs, Juan Somavía, Mahbub ul Haq, Enrique Iglesias, Rodolfo Stavenhagen, Maurice Strong, Shigeto Tsuru, and Samir Amin. At the conclusion of their meeting, these men issued a ringing manifesto:

> Thirty years have passed since the signing of the United Nations Charter launched the effort to establish a new international order. Today that order has reached a critical turning point. Its hopes of creating a better life for the whole human family have been largely frustrated. It has proved impossible to meet the "inner limits" of satisfying fundamental human needs. On the contrary, more people are hungry, sick, shelterless and illiterate today than when the United Nations was first set up.
>
> At the same time, new and unforeseen concerns have begun to darken the international prospects. Environmental degradation and the rising pressure on resources raise the question whether the "outer limits" of the planet's physical integrity may not be at risk.
>
> And to these preoccupations must be added the realization that the next thirty years will bring a doubling of world population. Another world on top of this, equal in numbers, demands, and hopes.
>
> But these critical pressures give no reason to despair of the human enterprise, provided we undertake the necessary changes. (Cocoyoc Declaration, 1974, pp. 170–1)

Cocoyoc brought together two major strands of the alternative movement: those who had argued that priority attention should be given to satisfying the "basic needs" of people for food, water, and shelter rather than to simple growth-maximization, and those who were concerned with the "outer limits" of the planet's resources and its environment to sustain such growth. A people-centered development in harmony with the environment would require a more self-reliant effort than in the past. Although autarchy is clearly not what they had in mind, the signers of the Cocoyoc Declaration thought that self-reliance might imply "a temporary detachment from the present system." They thought that "to develop self-reliance through full participation in a system that perpetuates economic dependence" would be impossible (1974, p. 174).

The following year, the Swedish Dag Hammarskjöld Foundation published a pamphlet with the provocative title "What Now? Another Development" (1975). Mainstream models of development and the policies based on them were challenged for failing to address the question of mass poverty and sustainability. In its ten-point program, the foundation advocated a humanist model of development

> of every man and woman – the whole man and woman – and not just the growth of things, which are merely means. Development geared to the satisfaction of needs beginning with the basic needs of the poor who constitute the world's majority; at the same time, development to ensure the humanization of man by the satisfaction of his needs for expression, creativity, conviviality, and for deciding his own destiny. (1975, p. 7)

In 1976 many of the participants in earlier meetings and projects, as well as some new contributors, came together in the establishment of the International Foundation for Development Alternatives (IFDA) in Nyon, Switzerland. Their principal purpose was to launch the Third System Project.

> We called this the "third system" not just by analogy with the Third World. The state and the market are the two main sources of power exercised over people. But people have an autonomous power, legitimately theirs. The "third system" is that part of the people which is reaching a critical consciousness of their role. It is not a party or an organization; it constitutes a movement of those free associations, citizens and militants, who perceive that the essence of history is the endless struggle by which people try to master their own destiny – the process of humanization of man. The third system includes groupings actively serving people's aim and interests, as well as political and cultural militants who, while not belonging directly to the grassroots, endeavour to express people's views and to join their struggle. This movement tries to assert itself in all spaces of decision making by putting pressure on the state and economic power and by organizing to expand the autonomous power of people.
>
> The two objectives of deepening and broadening the debate are indeed organically linked, for who are best able to look for alternatives if not those who suffer from the existing disorder and who need change? (IFDA, 1980, pp. 69–70)

In its progress report, "Building Blocks for an Alternative Development Strategy," the Third System Project took an innovative step by recognizing the different *scales* at which development occurs: local, national, global, and, somewhat ambiguously, "Third World." Of these, local space was regarded as the most significant for "people's creative unfolding": "Development is lived by people where they are, where they live, learn, work, love, play – and die. The primary community, whether geographical or organizational, is the immediate space open to most people. It is in the village, the neighbourhood, the town, the factory, the office, the school, the union's local, the party's branch, the parish, the sports club, the association – whatever its purpose – that personal and societal development first and best interact" (1980, p. 12).

But even as these words were being circulated to the 8,000 names on IFDA's mailing list (the list would grow to 25,000 by 1991!), the world was moving in quite another direction. Things were not going the Third System way. Though it was largely misunderstood in the West, China's Cultural Revolution with its austere egalitarianism had served as an inspiration for many advocates of an alternative development. But by 1979 China was setting out on a new course of modernization, accelerated growth, and mixed economy in which it would be acceptable, as China's leaders proclaimed, that a few should get rich even if the rest remained, temporarily at least, poor. That same year Margaret Thatcher was elected prime minister of Great Britain, and in 1980 Ronald Reagan was elected president of the United States. The neoconservative revolution was on its way.

Throughout much of the seventies, Western Europe and the United States pumped enormous amounts of money into the economies of what was then still called the Third World. These loans, a direct consequence of oil price increases set by the Organization of Petroleum Exporting Countries (OPEC), gave the receiving countries a fleeting illusion of prosperity. But when the loans began to be called in by uncharitable creditors, debtor countries discovered that to pay back even the interest on the loans they had only yesterday accepted would all but wreck their economies. Debt repayment and capital flight combined actually to create a net outflow of investment funds from poor to rich countries. In 1982 Mexico, the second largest of the debtor countries (with Brazil in the lead), declared itself unable to repay its loans, despite being a major oil producer. As other countries followed suit, the international debt crisis exploded. Per-capita indebtedness ranged from $200 to $2,000 in countries whose per-capita income was in the same range (Walton, 1989, p. 301). Clearly, countries such as Mexico, in which debt service as a proportion of exports was more than one-third, needed bailing out.

The principal agency acting on behalf of credit institutions in the West was the International Monetary Fund (IMF), which assumed the task of certifying debtor countries' continued credit-worthiness. As it turned out, certification was to cost dearly. To qualify for loans, countries would have to carry out "structural adjustment programs" geared to the requirements of a recently rediscovered neoliberal economic doctrine. Principal provisions were deregulation, the privatization of government-owned enterprise, elimination of tariff barriers to overseas imports, measures to promote export production, devaluation of national currencies to cheapen exports, and "getting markets to work properly" by removing subsidies and establishing the "correct" set of relative prices that would permit efficient, long-run growth (Griffin, 1989). In short, IMF-imposed austerity programs resurrected the laissez-faire dogma of nineteenth-century Manchester liberalism, with its blind faith in a self-regulating market. Keynesian interventionism was condemned because it was thought to undermine the natural dynamism of uncontrolled markets. Political and pragmatic considerations occasionally intervened to prevent dogma from being carried to its logical extreme. Even so, popular outrage at austerity measures, particularly the withdrawal of consumer subsidies, coupled with rampant inflation, high unemployment, and a decline in real wages, led to widespread urban unrest (Walton, 1989).

The neoconservative (counter)revolution was by no means simply willed by the political elites of Great Britain and the United States. Rather, it was the direct expression of a profound crisis within capitalism, which manifested itself in a vast perestroika of global dimensions. While old industrial areas were shutting down, new, "high-tech" industrial regions were booming (Storper and Walker, 1989). Some of these regions were in Mexico, along the border with the United States, and in the Philippines, Malaysia, Singapore, Korea, and Taiwan, where authoritarian regimes vouched for political stability, and where a highly disciplined and tractible labor force – especially young women – was available at a fraction of the price in the already industrialized countries. A new vocabulary entered the language: Pacific Rim economy, "four little dragons," transnational corporation, *maquiladora* (offshore assembly plant), new international division of labor, and newly industrializing country were among the choicer terms. Export-led growth, nudged by the IMF with the support of World Bank structural adjustment loans, was hailed as the new salvation.

In most countries, however, the combined effects of the debt crisis, structural adjustment, and desperate investments to promote production for export had disastrous consequences for both the poor and the environment. As real wages declined, in some cases by as much as 40 and 50 percent over the decade, pauperization became widespread. Export drives and the geopolitical strategies of military regimes contributed to the devastation of tropical rain forests, notably in the Brazilian Amazon (Hecht and Cockburn, 1989). Criminalized economies, based on coca crops and the production of cocaine and crack, threatened democracies in Bolivia and Colombia, where drug lords had suborned even the highest levels of government and were prepared to engage in terrorist acts to defend their empires (Castells and Laserna, 1989). And, increasingly in symbiosis with the drug economy, guerrilla movements flourished in Colombia and Peru.

The search for an alternative development could easily have been overwhelmed by these events. But it wasn't. In 1983 the United Nations General Assembly appointed the World Commission on Environment and Development "to propose long-term environmental strategies for achieving sustainable development by the year 2000 and beyond" (World Commission, 1987, p. ix). The publication of the Brundtland Report, so named after its chair, Norwegian prime minister Gro Harlem Brundtland, was hailed as a landmark and stimulated renewed discussion of alternatives, this time with a focus on sustainability. Environmental issues were becoming a global concern.

About the same time, some Third System people decided to push for an alternative development from outside established institutions. In 1984 they began to meet annually for what they called "The Other Economic Summit," at which they passionately debated such topics as "the international economic disorder," "putting people first," "in search of self reliance," "working like women," and "local economic regeneration" (Elkins, 1986). Surprisingly, and with a time lag of several years, their works were beginning to have some effect even on mainstream institutions, such as the World Bank, which by 1990 had put poverty back on the development agenda (World Bank, 1990) and had created special offices on the environment and women. But the by now familiar neoliberal doctrine was carried over, if in a slightly less aggressive form, into the new decade.

Assessment

Scattered experiences allow us to draw some tentative conclusions about widespread doctrinal beliefs about alternative development: (a) the belief that the state is part of the problem, and that an alternative development must as much as possible, proceed outside and perhaps even against the state (see Sanyal, forthcoming); (b) the belief that "the people" can do no wrong and that communities are inherently *gemeinschaftlich* (Campero, 1987); and (c) the belief that community action (i.e., action on the social terrain) is sufficient for the practice of an alternative development, and that political action is to be avoided.

Together, these beliefs spell out an all-but-anarchistic program along the lines of, say, Peter Kropotkin's writings (1970). The state is defined as the enemy. It is bureaucratic, corrupt, and unsympathetic to the needs of the poor. Often, it is in the hands of military and civilian elites who treat it as their private domain. Alternative projects are therefore frequently designed to bypass the state and to concentrate instead on local communities, which are considered moral and autonomous. People are said to possess ultimate wisdom about their own lives. For many "alternative" advocates, the voice of the people cannot be in conflict with itself; it speaks truly. And finally, given that the state is regarded as venal, politics is best avoided; it would only contaminate the purity of face-to-face encounters in neighborhood and village and thus (re)establish clientalistic relations that negate an authentic, people-centered development (Korten, 1990).

I do not share these beliefs. Although an alternative development must begin locally, it cannot end there. Like it or not, the state continues to be a major player. It may need to be made more accountable to poor people and more responsive to their claims. But without the state's collaboration, the lot of the poor cannot be significantly improved. Local empowering action requires a strong state.

Nor are communities necessarily *gemeinschaftlich*, even when they take part in a moral economy based on reciprocity and trust. Many fault lines run through both rural and urban communities: religious, ethnic, social class, caste, linguistic. And the universally subordinate role of women requires us to identify yet another source of social tension and conflict that cuts across all of the others. Each of the several social groups within a territorial community is likely to see its situation from its own perspective and contend over the same and always limited resources. Territorial communities are thus necessarily also political communities, rife with the potential for conflict.

Lastly, these conflicts cannot be contained locally. They are likely to spill over into regional and national political arenas. A politics of claiming is inherent in an alternative development, which is always about the use of *common* resources (usually controlled by the state) and the removal of those structural constraints that help to keep the poor poor. If an alternative development is to advocate the social empowerment of the poor, it must also advocate their political empowerment.

In the following chapters I raise these points repeatedly. Some may argue that I pay insufficient attention to conflicts even among the poor themselves. Poverty is merely one attribute among many. Some poor have access to land, others do not. Some are literate, others are not. The urban poor may have interests that conflict with those of the rural poor. Interest conflicts among the poor are indeed rife! My reply would be that this apparent lack of attention to conflict is a matter of emphasis more than blindness to social realities. A largely normative essay such as this cannot, at the same time, analyze actual political struggles in detail. My intention is to provide for alternative development what has so far been missing from the literature: an explicit theoretical framework outside the well-known neoclassical and Keynesian doctrines and thus a common point of departure for practice. Without such a framework, social learning, which is the connection between theory and practice, cannot take place. At appropriate points in my argument I therefore introduce a series of relatively unfamiliar concepts, such as domains of social practice, life space and economic space, the whole economy, the household economy, poverty as a form of social and political (dis)empowerment, and access to the bases of social power. These concepts acquire their specific meaning only within a politics of claiming that sets out a strong programmatic context for an alternative development.

In this perspective, which falls this side of revolutionary practice, alternative development does not appear as a game of winner take all. It can never triumph. It is not a complete alternative. It must be seen as the continuing struggle, in the long *durée* of history, for the moral claims of the disempowered poor against the existing hegemonic powers.

Moral Justification

Like the mainstream doctrine to which it stands in dialectical opposition, alternative development is not primarily a set of technical prescriptions but an ideology (Sutton, 1988). As such, it has a certain moral coherence. It also constitutes an immanent critique of the model in dominance. Under the sign of modernization, growth-maximizing development strategies contain an inherent and perhaps fatal contradiction. An early United Nations report put it quite bluntly:

> There is a sense in which rapid economic growth is impossible without painful readjustments. Ancient philosophies have to be scrapped; old social institutions have to disintegrate; bonds of caste, creed and race have to be burst; and large numbers of persons who cannot keep up with progress have to have their expectations of a comfortable life frustrated. Very few communities are willing to pay the full price of rapid economic progress. (United Nations, 1951, p. 15)

In many regions of the world, countervailing ideologies have contended this apocalyptic vision of economic development in the name of alternate sets of values, distinctive of the societies that gave them birth, principally India, China, and Middle Eastern Islamic countries. Rejecting Western models of modernization, these ideologies would help to write important chapters in twentieth-century history, including Gandhiism (Das, 1979; Sharma, 1987), Maoism (Gurley, 1976; Starr, 1979), and Islamic radicalism (Sivan, 1985). Although vastly different among themselves, they all rejected a system driven by relentless competition, forced to expand production continuously regardless of cost, while bringing ever-new technologies on the market. The nineteenth century called this progress; Joseph Schumpeter called it a "whirlwind of creative destruction" (1942; orig. 1934).

But proponents of an alternative development question the assertion that "creative destruction" is inextricably linked to the story of human progress. They demand that the question of what furthers human life be examined on its own merits. If social and economic development means anything at all, it must mean a clear improvement in the conditions of life and livelihood of ordinary people.[2] There is no intrinsic reason, moral or otherwise, why large numbers of people should be systematically excluded from development in this sense or, even worse, should become the unwitting victims of other people's progress. People have an equal and fundamental right to better conditions of life and livelihood.[3] The United Nations' experts, all of them economists educated at the very best schools in England and the United States, were wrong in pinpointing a group of persons who, they thought, "cannot keep up with progress," as if "keeping up" were a matter primarily of individual capacity. It is the very nature of technical and economic progress that excludes large numbers of people – in some countries a substantial majority – from its potential benefits. From the vantage point of the excluded, modern economic growth is not an intrinsic good, and the question of development cannot be left exclusively to those for whom the summum bonum is found in market relations and growth maximization. The human and environmental costs of economic growth must be considered as well.

Taken by itself, alternative development cannot, of course, be the complete story. As an ideology, it argues for the rectification of existing imbalances in social, economic, and political power. Centered on people rather than profits, it faces a profit-driven development as its dialectical other. *Actual* development will always be the historical outcome of the ideological and political conflicts between them.

---

[2] This formulation owes a great deal to Karl Polanyi, who insisted on the importance of what he called the substantive economics of livelihood, considered as practices deeply embedded in the matrix of social and cultural institutions (Polanyi, 1977; Stanfield, 1986).

[3] It would seem difficult to dispute this claim. Still, some argue that development is inherently unequal and that the benefits of development will "trickle down" in time to those initially bypassed. But that promise is reassuring only to those who already share in the new bounties. For the rest, it remains an uncertain promise.

What are the grounds for preferring an alternative to long-term growth maximization, which is the bottom line in mainstream theory and practice? The question has more to do with morality than facts. What is the basis for claiming that every person is entitled both to adequate material conditions of life and to be a politically active subject in his or her own community? I shall argue that such claims have three foundations: human rights, citizen rights, and "human flourishing."

The Universal Declaration of Human Rights was adopted by the General Assembly of the United Nations in 1948. According to legal scholar Beth Andrus (1988),

> What began as a General Assembly resolution has evolved into a set of customary international legal principles. Although this view was first advanced solely by legal scholars, it has achieved widespread acceptance by governments assembled at international conferences, by state practice, and by court decision. Not only has the world embraced the rights and obligations set forth in the Declaration, but the Declaration has served as a seed from which an entire system of protection has grown. This system has innumerable branches all springing from a common purpose: the promotion and protection of international human rights. (pp. 4–5)

But the declaration does something more profound than establishing a legal system of protection. It sets forth a universal code of basic moral conduct to govern our relations with each other as social beings.

Articles 3–21 of the declaration identify civil and political rights, or the right to liberty. Economic and social rights follow (Articles 22–27). Although the United States has chosen to give more weight to civil liberties than welfare rights, the latter, as Susan Moller Okin argues, are just as fundamental as those of the first kind (1981). Article 25.1 states: "Everyone has the right to a standard of living for the health and well-being of himself and his family, including food, clothing, housing, and medical care and necessary social services, and the right to security in the event of unemployment, sickness, disability, widowhood, old age, or other lack of livelihood in circumstances beyond his control" (1981).

Willful exclusion from these rights is a kind of violence on the person excluded. A right tells us when a wrong is being committed; it also tells us what norms ought to prevail. This is not to deny that there are practical problems in spelling out the meaning of a "standard of living" and of "security." Such problems can be worked on. The assertion of a human right tells us that there is a grave problem that must be addressed.[4]

---

[4]   For an application of human rights to the housing question, see Leckie (1989).

My second argument concerns citizen rights, and it is more straight-forward. Citizen rights are specific to states and to conditions prevailing therein. They presume nothing more (but also nothing less!) than the formal recognition of citizen status – that is, of citizens' relative auto-nomy vis-à-vis the state. They presume, therefore, a modern, democratic state, where the holders of authority are ultimately accountable to the people organized as a political community. In such a state, the rights enjoyed by one group of citizens must be equally granted to all. Citizen-ship is categorical: one cannot be half a citizen. The rights and duties of citizenship fall equally on all who can rightfully claim membership in a political community. As Michael Walzer put it, "The primary good that we distribute to one another is membership in some human com-munity. And what we do with regard to membership structures all our other distributive choices: it determines with whom we make those choices, from whom we require obedience and collect taxes, to whom we allocate goods and services" (1983, p. 31).

In any given community, some people – for example, women in Switzerland (until quite recently), Qechua-speaking Indians in Peru, Arabs in Israel, the "underclass" in the United States – may indeed be excluded from the *practice* of full citizenship. But their historical strug-gles have been to gain an *equality* of rights with citizens already con-firmed in their rights and thus politically empowered (Hardoy and Satterthwaite, 1989). Their claim is for full citizenship de facto.

Two groups commonly excluded from the claims of citizen rights are children and future generations. Their claims derive from the observation that every human community, and specifically every political community, seeks to perpetuate itself. It thus exists in time as well as space; it has a collective past and desires a future for itself. It cannot be argued without contradiction, therefore, that future citizens (who cannot speak for them-selves but who may have living advocates) are entitled to fewer rights (or perhaps to none at all) than those claimed by the present generation of adults. A basic fellowship binds citizens of all generations. It is that interest in a common life, whether in defense against hostile forces from without or divisive forces from within, that constitutes the bedrock of citizen rights. To be excluded from them is to be deprived of a primary good.

The third ground for an alternative development is what Margaret Jane Radin (1987) has called "human flourishing." The term is evocative and open-ended. It asks us to consider what it means to be a full human being living up to her or his capacity. It argues for the right to those social conditions, both general and specific, which make a human flourishing possible. We may never know the ultimate limits of such flour-ishing, nor may we be certain of the conditions that are most conducive to a proper human development. But we can and do know what *inhibits* it: hunger, poor health, poor education, a life of backbreaking labor, a constant fear of dispossession, and chaotic social relations. Their removal will encourage human flourishing. An alternative development is therefore justified because, by improving the conditions of life and livelihood of the excluded majorities, it promises to further the self-development of human beings as individuals.

As in many places throughout this study, the question can legitimately be raised whether the values propounded – in this case human flourishing, with its deep roots in the Judeo-Christian tradition – is appropriate as a normative basis for an alternative development. Is it not true, a critic might ask, that an alternative development must always be culturally specific, and that it is precisely this specificity that distinguishes it from mainstream doctrine? Shouldn't alternative development always be expressed in the plural form, as each culture, each people, each ethno-regional group lays out its own distinctive path to the future? People living within the Confucian tradition, for example, may not regard human flourishing as a primary good. In the Confucian world of China and Korea, the primary good is the patriarchal family to whose collective well-being individual members are expected to subordinate their own wishes and desires. And there are many other examples of this sort.

I would answer such critic with both yes and no. Yes, ideally, an alternative development is always specific to the collectivity whose future is in question. Maoism, to cite one example, charted precisely a Chinese mode of development, different from that of the capitalist West. But to bury alternative development in the language of cultural relativism and endogeneity (that all development must germinate from within a particular culture) would be to silence all development discourse while giving free rein to the existing hegemonic system, which is fuelled by Western ideas and ideals and is wholly untroubled by questions of cultural relativism. It is one thing to argue for an alternative development in which "human rights" and "human flourishing" are major assertions of value, but quite another to impose these values on any group of people against their will. The alternative development in this study is a Euro-American construct, but it has found supporters in many parts of the the world outside its region of origin. As we move toward a single, interdependent world economy, it may become a major counter force to the dominant system of accumulation.

The alternative development of this essay accepts the existing system of global accumulation as a fact. It does not propose to turn away from it and shut the door: the isolation and inward orientation chosen by a Burma is neither feasible nor desirable for the rest of the world. There are other options than to shut oneself off from a global system whose existence in no way proves its immutability. No matter how dynamic, an economic system that has little or no use for better than half of the world's population can and must be radically transformed. Broadly speaking, the objective of an alternative development is to humanize a system that has shut them out, and to accomplish this through forms of everyday resistance and political struggle that insist on the rights of the excluded population as human beings, as citizens, and as persons intent on realizing their loving and creative powers within. Its central objective is their inclusion in a restructured system that does not make them redundant. It is a moot point whether capitalism so transformed can still be called capitalism. Given the many historical varieties of capitalism, it is perhaps an unimportant question.

2   The Trajectory: From Exclusion to Empowerment

Economic and Political Exclusion

The recent convulsive changes in the organization of capitalism – its global reach, its revolutionary technological innovations, its centralization in giant corporations and financial institutions – have resulted in the virtual exclusion of vast numbers of the world's poor from effective economic and political participation.[1] The message sent to them is very clear: for all practical purposes, they have become largely redundant for global capital accumulation. Modern capitalism can, for the most part, do without subsistence peasants, landless rural workers, and the so-called popular sectors of rapidly growing urban slums and shantytowns. Some, in fact, perceive them to have negative effects on capital accumulation, on the grounds that the urban poor siphon off capital for relatively unproductive public expenditures on housing, education, and health, and that subsistence peasants obstruct necessary modernization in agricultural production. Multinational corporations need land to organize the production of industrial and export crops, and small peasants occupy

[1] The argument about the exclusion of those living in the subsistence sectors is a complex and contentious one. Because virtually all people living at subsistence levels participate in the market, the argument concerning their "exclusion" would seem to carry no force. Marxists, for example, insist that the popular urban sector – constantly replenished by rural migrants – constitute a "reserve army" of labor whose chief function is to keep wages in the formal or accumulation sector as low as possible. Although these arguments have some merit, the disproportion between the actual number of the excluded population and their effective purchasing power is so great that the contribution of the excluded population to the processes of capital accumulation is at best marginal. Much of what they buy is, in fact, produced within the subsistence economy and has only tenuous links with the channels of accumulation. As for the reserve-army argument, it is of course true that a relative surplus of labor will lower the price of labor. The question to consider is how much of a surplus is necessary to bring wages down to their actual levels in poor but labor-rich economies. Is a 10 percent surplus sufficient? Is 20 percent? Whatever the figure – and there unfortunately are no firm data to back this argument – it is unlikely that a majority of the economically active population, including small-farm peasants, landless workers, and the popular sectors in cities, is needed to produce the actual wage levels in the unskilled occupations. A very large portion of the work force, therefore, would, from the perspective of capitalist accumulation, appear to be largely redundant.

the coveted land. Landless rural workers with a capacity to make wage demands pose a threat to capitalist producers, and the urban popular sectors take their noisy protests into the street. Although the popular sectors provide many useful services at very low cost, they are increasingly regarded, as were their nineteenth-century counterparts in Europe, as the "dangerous classes" that oblige the state to maintain a large, if unproductive, security apparatus.[2]

How does this massive redundancy come about? Figure 2.1 helps us to find an answer by presenting a model of a poor national economy in which a substantial part of the population makes its living in rural areas.[3] The diagram is divided into a capital-accumulation sector, with about 40 percent of the country's population, and a subsistence sector. The latter consists of three categories of peasant: small peasants with access to land; landless peasants who depend mostly on wage work for their subsistence even though they still reside in rural areas; and a category of peasant farmers who, in addition to producing for their own subsistence, grow marketable crops as well, some of them for export. Commercial farmers and agribusiness predominate on the capital-accumulation side of the diagram. The former are still engaged in family farming but employ hired labor for a significant part of the work, whereas agribusiness relies exclusively on hired labor and uses increasingly capital-intensive methods of production. Both forms of capitalist production are closely linked to multinational concerns that, inter alia, provide management and marketing services as well as basic inputs, such as hybrid seeds, for the production process (contract farming). Constituting a transitional category, peasant farmers in the subsistence sector may become similarly dependent on multinationals.[4]

The diagram shows the urban-metropolitan economy as the focal point of economic relations. As is proper, the metropolis straddles both the capital-accumulation and subsistence sectors. In a rough-and-ready identification, the urban subsistence sector is also known as the informal economy.[5]

Commodity relations dominate transactions in the accumulation

---

[2] In Latin America the term *popular classes* is broadly applied to the urban working class, irrespective of their insertion into the market economy or their relation to the ownership of the means of production. Many covered by this term are self-employed; most women work in the domestic sphere; the majority make their living in informally organized work; and they are often recent immigrants to the city. The popular classes or "sectors," of course, exhibit internal social divisions and potential for conflict, but these are not at issue here.

[3] In 1987 low-income countries called 70 percent of their population rural, followed by 49 percent in lower-middle-income countries and 33 percent in upper-middle-income countries (World Bank, 1989a).

[4] For the concept of peasant, see Shanin, 1987, pp. 1–12.

[5] For further discussion of "informality" in economic relations, see chapter 5.

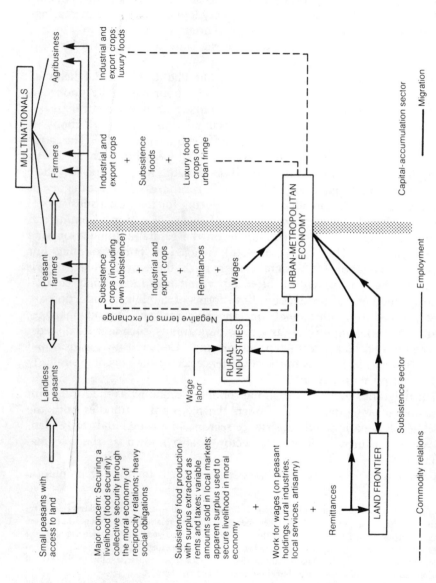

Figure 2.1 The transformation of the peasantry in capitalist social formations

sector. In the subsistence economy, they are supplemented by what some have called the moral economy or economy of affection (Scott, 1976; Hyden, 1980).[6] Operating largely outside the market framework but articulated with it, the moral economy is of critical importance to the livelihood of both small peasants and the urban poor. It is based on the principle of reciprocal exchange among kin, near-kin, neighbors, and friends, and it governs clientalistic relations to patrons.[7] Notwithstanding the importance of nonmarket relations in the survival economy, money continues to be required by even the poorest. Small peasants in particular, though producing primarily for their own use, are obliged to sell part of what they produce to purchase essential commodities and to fulfill their social obligations, which, in turn, gives them a measure of security in the moral economy.

Small peasants are typically settled in land-short areas. As their numbers grow, poverty eventually forces many of them to abandon their homesteads. Some become landless workers, moving to towns and cities near their ancestral villages. Others undertake longer journeys either to the land frontier (if there is one), where they tentatively reestablish themselves on cleared forest land, or to metropolitan areas where they are absorbed into the informal economy. But peasant farmers may also find their subsistence needs "squeezed" (i.e., the prices of what they produce tend to decline relative to the prices of urban-made products) and may prefer to sell or rent their land to farmers whose economic situation is better than their own. Some peasant farmers will end up as landless workers in rural districts, but a few will make it across the great divide into the accumulation sector of the economy, becoming full-fledged capitalist farmers. The remainder move to the metropolis.

The sources of small peasants' and peasant farmers' monetary income are various; they include the selling of surplus staples as well as industrial and export crops, frequently at the expense of their own food security. Both groups derive income from nonfarm work in rural occupations, as well as from remittances sent by family members who have gone to work in the city. Finally, as shown by arrows at the top of the diagram, both may engage in agricultural work for wages.

In relative numbers, the three categories of the rural subsistence

---

[6] The moral economy school derives its theoretical inspiration from the writings of Karl Polanyi (1977); see also Stanfield, 1986.

[7] Following Lévi-Strauss and others, Peter Ekeh (1974) distinguishes between mutual (dyadic) and generalized (univocal) reciprocity. The former involves short-term exchanges of goods and services of approximately equivalent value between two parties. They may be either horizontal exchanges or vertical, as in clientalistic relations. Generalized reciprocity refers to the social obligations of an individual or small group toward larger collectivities of which they form a part (e.g., families, clans) and on whom their own continued well-being depends.

population are increasing at only about half the rate of the national population, while urban population is growing from 1.5 to 2.5 times the national rate. The landless (or near-landless) population is expanding faster than the two "landed" groups, receiving influx from both, and in some countries already comprise upwards of 40 percent of rural population. Over time, most small peasants become absorbed into the so-called popular classes, but peasant farmers may survive for a long time, though in sharply diminishing numbers.

As we turn to consider the capital-accumulation sector in the diagram, we find both farmers and agribusiness relying heavily on waged labor. Emphasis in production is on crops for industrial use and export, such as sugarcane, cotton, coffee, and tea; truck gardening for urban markets; and marketable staples. Relations with the city are entirely of the commodity form. Being relatively capital-intensive, both categories of producers depend heavily on such metropolitan-economy inputs as machinery, fertilizers, research, management, and information.

Figure 2.1 shows the general trend of rural/urban transformation under peripheral capitalism: the increasing commercialization of agriculture, with its growing reliance on export-based production; the proletarianization of agricultural labor; the swelling numbers of the marginally employed in urban-metropolitan regions; the rising influence of multinationals in the management of agricultural resources; and the gradual disappearance of the land frontier (with accompanying problems arising from deforestation, the ethnocide of indigenous populations, and the indiscriminate destruction of wildlife).

These tendencies, which are to some extent present in all parts of the impoverished world, can be ascribed directly to the application of the mainstream model of economic growth. Capital from both domestic and international sources is invested primarily in the accumulation sector of the economy. Except during the intense drama of civil war and famine, the subsistence economy remains largely invisible. At best, it receives symbolic attention from international aid programs that offer small amounts of money on programs for the poorest. Some economists call them "tail-end poverty" programs. They are mounted primarily for reasons of charity, welfare, and political expediency.

Given this scenario, it is not especially surprising that statistics on income (not to speak of wealth) distribution are hard to find. Out of 96 low- and middle-income countries reported by the World Bank (1990, table 30), only 21 have benchmark data on who receives what shares of income. If we exclude the 42 poorest countries with per-capita annual incomes of less than $500 on the basis that available data are of doubtful validity, we are left with 54 middle-income countries and distribution data for only 15. Of these, three countries – Poland, Hungary, and Yugoslavia – fall outside our immediate concern. For the remaining countries, the lower 60 percent of the households receive the following proportions of national income: less than 20 percent, three countries; 25 to 29 percent, eight countries; 30 percent or more, one country.

The worst case on record is that of Brazil. In 1983, 60 percent of Brazilian households received less than 20 percent of the income, while the top 10 percent garnered nearly one-half. From figures such as these, it is clear that vast numbers of people live on the margins of the money economy, sustaining their livelihood in ways that rely heavily on work performed outside the money economy. By comparison, in 13 OECD countries the same 60 percent of households received between 32 and 40 percent of national income, with Belgium and Japan having the most equitable distributions.

Another way of showing the dismal meaning of these figures is by a hypothetical example. Assume a lower-middle-income country, such as Thailand, with a per-capita gross national product (GNP) of $1,000. (For comparison, U.S. per-capita GNP is nearly $20,000.) Ten percent of its households receive 45 percent of the income, or $450 million. This yields a per-capita income of $4,500 and, assuming a household size of 5.5 members, a respectable household income of $24,750.

The 40 percent of the poorest households, on the other hand, receive only 10 percent of the income (as, for instance, in Brazil), or $100 million, for a per-capita figure of $250, and an annual household income, by the same assumption as before, of $1,375. The income of the privileged 10 percent is thus 18 times greater than that of the poorest two-fifths. In other words, whereas the upper 10 percent clearly participate in capital accumulation and are appropriately rewarded, the lower 40 percent participate only marginally.

The overall dimensions of this brutal exclusion from "development" must be kept in mind. More than three billion, or two-thirds, of the world's population have incomes that are less than 10 percent of the per-capita income in the United States. A short generation from now, by 2025, their numbers are expected to double, rising to more than 70 percent of projected world population (World Bank, 1990). Per capita income, of course, is an inadequate measure of relative poverty, which is a multi-dimensional phenomenon whose definition cannot be separated from underlying ideological considerations. Still, the income measure can give us a rough order of magnitude of the extent of global poverty, becomes apparent when entire countries are taken as the unit of measurement. In addition, within each country, whether rich or poor, large numbers of people live at or below a culturally defined level of subsistence and minimally contribute to the GNP. The presence of this massive "underclass" at both relevant scales – global and national – is the principal reason for thinking about an alternative development.[8]

To be economically excluded is, for all practical purposes, to be politically excluded. Contrary to a good deal of earlier thinking, political development toward a more inclusive democracy is neither the inevitable complement of economic growth nor its predestined outcome. On the contrary, in most parts of the world today, the unleashing of capitalist accumulation does not occur in the benign climate of a liberal democracy. So-called modernization regimes, whether in Brazil during the late sixties and seventies or in Chile during the seventies and eighties, are typically run by strongman presidents, military juntas, and technical wizards (Falk 1980). Political silence is enforced against the "invisible" population of the excluded sectors and their political champions by jailings, torture, and "disappearances." The rule of the modernizers is largely by fear.

But state violence cannot silence resistance forever, and the inevitable happens: corruption flourishes, policies misfire, privileges are abused. In the end even the most powerful of dictators overreaches himself, until he is finally forced to relinquish office. Some deposed leaders, have sought exile in the West, some have died, others have declined into comfortable obscurity. A period of democratization follows their demise.

With democratization come new challenges and dilemmas. Impoverished sectors can at long last voice their demands, and political opposition returns, often chaotically, with dozens of new and resurrected parties contending for the voter's attention. At the same time, the errors, mismanagement, and corruption of the past continue to plague the new authorities. Huge amounts of export earnings have to be paid to foreign

---

[8]   The World Bank (1990, chapter 2) arrives at a much lower estimate of what it calls poverty by adopting global poverty lines of, respectively, $275 and $370 per person per year, which are typical of poverty lines used in countries such as Bangladesh, India, Tanzania, and Egypt. On this procedure, the bank arrives at a 1985 global estimate of 633 million and 1,116 million, respectively for each of the two poverty lines. Their count thus includes only the poorest of the poor – a useful concept perhaps, but different from the one used in the present study.

According to a news item in the Los Angeles Times for July 14, 1990, the United Nations Economic Commission for Latin America has estimated that 64 percent of the region's population (including all of Latin America and the Carribean) lived in poverty, with 20 percent living in absolute poverty, meaning that they do not have enough to buy the minimum food for a basic diet. This estimate roughly coincides with the claim made in this study that a majority of the world's population – perhaps as much as two-thirds – are in some sense poor. For an exhaustive descriptive account of evolving poverty in Africa, see Iliffe, 1987.

[9]   Countries with mobilization regimes similar to Brazil's and Korea's in 1980 include Argentina, Chile, Taiwan, Egypt, Indonesia, Iran, Mexico, Nigeria, Panama, Peru, the Philippines, Saudi Arabia, Singapore, South Africa, Thailand, and Uruguay (Falk, 1987, p. 112).

creditors.[10] Debt service has to be renegotiated. But even as people are clamoring for more equitable growth, the newly minted democratic regimes are obliged, as a condition of further international assistance, to adopt "structural adjustment" policies including trade liberalization, the elimination of consumer subsidies, privatization, currency devaluation, and favorable treatment for multinational concerns. None of these policies are designed to bring "development" to the popular sectors. Desperate attempts by the government to satisfy popular demands fuel hyperinflation, reducing the chances for a sustained development still further. Street demonstrations are quelled with the usual violence. Gradually the military recovers its traditional role as enforcer, and the groundwork is laid for yet another cycle of militarization, technocracy, and repression.

This vicious cycle has characterized Latin American politics (but not only Latin American politics!) for half a century (O'Donnell and Schmitter 1986). There is no evidence that it is about to be broken. A good example is the Philippines under President Corazon Aquino, who has had to fight off a series of attempted military coups since the democratic people's revolution brought her to power in 1986. Argentina appears to be undergoing a similar process. These cycles, of course, do not obey some inevitable logic of history. The practices of an inclusive democracy can be learned, albeit painfully. Opposition to mainstream economic policies is not entirely without response. And with each new cycle, civil society is gaining strength.

Growth-maximizing strategies have isolated the poor and in many respects have made their lot worse. But the success of exclusionary policies, even when measured by the narrowest of criteria has been limited. Many countries that only a decade ago were touted as becoming "newly industrialized" suffer today from some combination of hyperinflation, rising foreign indebtedness, unbalanced economic growth, civil disorder, and environmental pressures. Some, of course, have made relatively successful transitions, most notably South Korea, Taiwan, and Malaysia. But in more than one country that has experienced a measure of democratization, the return of military rule is an ever-present threat. It is also true that throughout their histories, many countries have never known democracy. With the exception of India, the poorest countries of the world remain under "praetorian" rule.[11]

---

[10]   As a percentage of export, debt service in 1987 stood at 27 percent for Brazil, 22 percent for South Korea, 30 percent for Mexico, 45 percent for Argentina, and similar orders of magnitude for the remaining countries on Falk's list of authoritarian modernizing regimes. See World Bank, 1989a, table 24.

[11]   According to Falk (1980), "Praetorian rule . . . involves the seizure and exercise of state power by an elite that does not rest its legitimacy on any system of political accountability to its citizenry. Furthermore, recourse to autocratic rule is not a consequence of a central economic

Everyday Resistance and the Dialectic of Modernization

To be excluded from the upper echelons of power is not necessarily to be condemned to a life without history (Wolf, 1982). The will to resistance is formed in everyday struggles for survival (Scott, 1985, 1990). Against the dismissive image of the underclass projected on its members by the ruling strata of society – that the poor are indolent, untrustworthy, improvident – a new identity, found in the very acts of collective resistance, is being asserted (Evers, 1985). To the rich, the poor may be largely invisible. To each other they are known in the dialogue of daily encounter.

Central to the struggle for survival is the household economy, the moral economy in which households are imbricated, and the territorial communities that form the matrix of household, social and political practice. In the extreme case, these supports may be missing. But for the majority of the poor, to be excluded does not mean to be discarded. People find strength in solidarity with others much like themselves (Lomnitz, 1977).

The struggle for survival can take many forms. Some are acts of individual enterprise in the informal economy; some are collective acts of protest and defiance; still others are centered in the community. All initiatives require the cooperation of others; most require some form of outside help from students, priests, and professionals who may provide the catalytic spark in the face of adversity. By making common cause with the poor (and sometimes even sharing their lot), these volunteers are able to strengthen resolves, provide technical counsel, and mediate with outside agencies and the state.

That the daily struggle for survival among the lower strata of the excluded sectors is also an act of resistance is relatively recent insight. James C. Scott (1985), writing of Malaysian peasants, called it "the weapons of the weak." I want to cite a few examples from urban Latin America to show the living context of such resistance and its deeper significance.[12]

Informal work

Informality is broadly defined as work that is not controlled by the state and that takes place, so to speak, without the state's knowledge. The

crisis arising out of the need to deepen capitalist development or to discipline labor demands for economic shares and rights of participation. In the praetorianized instances, autocratic leadership may seek to vindicate its authority externally by a claim to provide sufficient order to assure profitable investment and steady economic growth" (pp. 86–7).

[12] For the general character of urban social movements in Latin America, see Ballón, 1986; Campero, 1987; Calderón, 1986; Kowarick, 1988; and Jacobi, 1989.

work may be legal or illegal. The former includes domestic work, street vending, carrying loads, preparing food for sale in open-air stalls, plumbing and electrical repairs, small-scale manufacturing, and construction. The latter includes gambling, drug selling, prostitution. In a city like Lima, Peru, most people work informally (De Soto, 1989). Some are tied into formal businesses through the expedient of subcontracting; many are not, and each day must hustle for a living. Perceived as a public nuisance, they may be periodically harassed by the police. Self-built housing from "found" materials are bulldozed in clearing operations because land to which the builder has no formal claim is said to be "invaded." Illegal workers have an even harder time. Yet despite this constant harassment, the number of informal workers continues to swell. Their persistence in the face of a hostile state is an act of civil resistance.

Popular economic organizations (PEOs)

So called by Chilean sociologists, PEOs refer to cooperative activities centered in the community whose immediate aims are either to lower the costs of household subsistence or to provide extra income through cooperative work. An example of the first are the *ollas comunes* (communal kitchens). Within a given barrio women get together to collectively prepare one hot meal a day for the members of the *olla* (literally, a pot). Ingredients – vegetables, rice, meat – are bought collectively (*comprando juntos*) or obtained from relief organizations, and meals are sold at cost. Labor is contributed free of charge. In addition to solving a major problem of malnutrition, especially among children, participation in an *olla* also teaches about the values of organization, joint decision-making, and leadership (Hardy, 1986).

An example of the second kind are the artesanal workshops (*talleres*) in Santiago that produce goods for the market. As in the previous example, participants are often women. Initially, they are organized with outside help, such as a Catholic action group. As women put their own productive talents to work, perhaps to make knitted goods, they are taught new designs, the importance of quality control, marketing skills, and bookkeeping. Some capital assistance may also be provided. The work may be done together in a member's spare room, or each woman takes her work home. Important are the periodic meetings of the *taller* where common problems are discussed and decisions made. After paying for materials and covering other expenses, proceeds from work are equitably distributed among *taller* members according to an agreed-on formula (Hardy, 1984).

In the case of popular economic organizations with their base in local communities, resistance is far less confrontational than with informal work. Rather, it lies in the discovery by participants of the powers of mutual aid and cooperation on the fringes of the market economy. Popular economic organizations may bring only a small amount of income, or they may simply succeed in lowering living expenses. Their major contribution is that they bring hope, teach skills, and, by turning individual problems into collective ones, offer new possibilities for solution. They are also essential vehicles for self-empowerment (Sabatini, 1989).

When interviewed as to why they participate in a *taller*, 40 percent of the respondents gave conviviality as their reason: the human dimension of sharing was identified as the single most important benefit. Another 30 percent gave reasons related to their status as women: to escape domestic drudgery, to acquire a new sense of personal dignity. And 28 percent gave personal growth as reason, such as the acquisition of new skills and useful information and the discovery of a new sense of self (Hardy, 1986, p. 113). Economics, of course, is at the base of both the *ollas* and *talleres*. It is what ultimately justifies the work. But the reasons of sociability, liberation from women's subaltern condition, and personal growth are more salient in the minds of the participants.

Protest movements

These are the best-known form of civil resistance. They invariably address collective needs such as housing, water, sewage disposal, and health services, and their form of protest is usually filled with drama. To succeed in their endeavors, protest movements require leadership, time, commitment, and above all an organizational framework to carry on with the struggle over the necessary stretch of time, which may be many years. Demand-making struggles bring residents into direct encounter with state authorities, whose response is often unfriendly at first. The agencies involved will initially try to meet the challenge by ignoring popular demands, then by making false promises, engaging in symbolic but essentially meaningless action, attempting to divide community leadership, and using physical force to subdue the movement. But if residents persist in their demands, their struggle will often be at least partially successful: they will continue to live in self-built housing on contested sites, and public services will begin to be provided (Jacobi, 1989; Eckstein, 1988; Pezzoli, 1990).

Protest movements tend to die out once people are reasonably satisfied that their demands have been met. But in a few cases, what begins as a protest movement in defiance of state authority may be transformed into an instance of self-management and become part of the fabric of a new society. A case in point is Villa El Salvador outside of Lima. Its story may be unique, but it is also instructive (Ballón, 1989).

It all began with an illegal invasion of terrain on the outskirts of Lima in 1971. The settlement was soon relocated to a prepared terrain in the desert by SINAMOS, the state agency responsible for social mobilization – that is, for bringing the working-class population of Lima under state control. A communal organization was set up to channel people's demands to the state, but CUAVES (the Self-Managed Community of Villa El Salvador) soon made itself independent of state sponsorship and remains autonomous to this day. Meanwhile, the population of Villa El Salvador has grown to a quarter million. In 1983 the settlement was incorporated as a municipal district with its own government. The material living conditions of the district are similar to those of other popular settlements in the poverty belt surrounding the capital of Peru. Average family income for a family of six is $70, and one-fifth of the households receive less than $30. Open unemployment and underemployment exceed 60 percent of the work force. Seventy-one percent of the population is under 25 years of age.

The difference is that Villa El Salvador encourages active citizen involvement. This has led to a highly successful process of open planning that directs the scarce available resources toward purposes that serve the community as a whole. Community organization is both territorial and functional. Territorial organization begins with households, 24 of which constitute a block, with 16 blocks per residential group and 22 groups for each sector. Representatives from each of these levels elect the board of CUAVES, which, while remaining a community-based organization, speaks with authority for the entire district. Functional organizations include communal kitchens, women's clubs, popular lending libraries, youth organizations, sports clubs, and so forth, each of which seeks to legitimize itself by the degree to which it is representative of its territorial base – the block, the residential group, the sector, and so forth. In addition, three large, districtwide federations have been formed: of women (9,000 members), small industrialists (600 members), and small retail merchants (600 members).

These organizations, totalling 4,000 units, participate directly in districtwide development planning. They also engage in their own demand-making, rights-claiming struggles. And from these they have learned that claiming rights is not enough; success comes when their own energies are also put on the line. As Eduardo Ballón (1989) tells it, "They have discovered the need to combine demands with their own actions (*gestión*). In many cases, this has led to conflicts over specific issues, but, paradoxically, these conflicts have encouraged greater social cohesion and, amidst a multiplicity of social practices and diverse organizations, have prevented a decline into social atomization" (p. 7). In other words, within a democratic framework, conflict is capable of producing solidarity and strengthening social cohesion. It also gives people a pride of local citizenship: they are *vecinos* (neighbors) of Villa El Salvador.

There is a great deal more to the story, but the evidence is clear: Villa El Salvador stands for a new reality in poverty-stricken Peru; it is an autochthonous territorial organization, with crisscrossing functional organizations, based on an inclusive democracy that has survived for more than 20 years and constitutes one of the largest urban *municipios* in Peru. It is a story of conflict, resistance, and cooperation, idealistic in its vision of a self-managing society and pragmatic in its search for solutions. In a Peru whose economy is on the point of collapse, it cannot lift the burden of poverty by itself. But it can create the conditions for an alternative development: an inclusive democracy that gives voice to those who are usually silent and the deployment of common resources to address the collective needs of a population that has been left largely to its own devices.

The resistance struggles of the excluded barrio residents of Latin America open up new vistas of existential reason. Whereas modernization in the upper echelons of society emphasizes materialism (the consumer society, technical progress), objective science, individualism, and liberal democracy, the ideology that comes out of the lived experience of the barrio emphasizes (a) intersubjective solidarity relations based on trust, (b) a reality testing based on subjective experience and intersubjective validation, (c) an anthropology of personhood, and (d) a political order based on the "strong talk" of direct democracy (Friedmann, 1989a).

Neither the existential reason of the barrio nor the cognitive reason of the Enlightenment implicit in the ideology of modernization is of itself sufficient and complete. The whole results from complementarity and conflict between them. But this dialectical "unity of opposites" is prevented from occurring as long as most of the population continues to be excluded from effective economic and political power. An alternative development must base itself on the existential rationality of the barrio. It must also seek to join the processes of existential reasoning with those of cognitive reasoning. Dialectical unity is achieved through the interpenetration and mutual transformation of the two.

Domains of Social Practice

To make sense of these stories of resistance, we need a conceptual map of the popular-struggle terrain. Terms like *state* and *civil society* require a holistic mapping that show them in relations of power. Such a mapping is attempted in figure 2.2, which shows four overlapping domains of social practice: the state, civil society, the corporate economy, and the

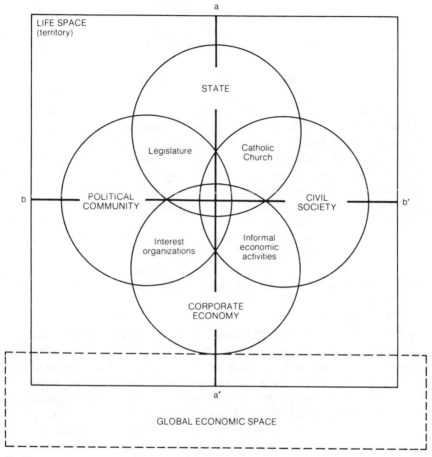

LIFE SPACE
(territory)

a

STATE

Legislature   Catholic
                     Church

POLITICAL   CIVIL
COMMUNITY   SOCIETY

b   b'

Interest   Informal
organizations   economic
                     activities

CORPORATE
ECONOMY

a'

GLOBAL ECONOMIC SPACE

STATE: state power; executive and judicial branches

CIVIL SOCIETY: social power; natural persons, households, and civil associations
              (the domain of culture and social structures)

CORPORATE ECONOMY: economic power; corporations and financial institutions (ie., judicial persons)

POLITICAL COMMUNITY: political power; social movements and political organizations

Figure 2.2 The four domains of social practice

political community. Each domain has an autonomous core of institutions that governs its respective sphere. The core of the state consists of its executive and judicial institutions; the core of civil society is the household; the core of the corporate economy is the corporation; and the core of the political community is independent political organizations and social movements. For each of these a distinctive form of power can be identified – state power, social power, economic power, and political power – according to the resources that actors in the domain can mobilize.

Our domains are highly schematic of course. When we look more closely their apparent unity breaks down, and their clear boundaries begin to merge. The state has many centers of power that are frequently at odds with each other and rarely act in a coordinated way. Civil society is divided along familiar lines of social class, caste, ethnicity, race, religion, and gender. The corporate economy consists of actors who are in fierce conflict among themselves, combining only when it furthers their collective interest. And the political community is, of course, the quintessential terrain of conflict fought out among different groups and factions involving the three remaining domains. Moreover, each domain exists at different levels of territorial organization: city, region, nation, and even multinational unions such as the European Community. Evidently the interests of each domain at each level of territorial organization are unlikely to coincide precisely, although there may be room for coalition building around specific issues.

The diagram is nevertheless a useful starting point for understanding the conflicts in territorial communities. It maps a sociopolitical space that allows us to locate specific institutions such as legislatures (between state and political community) or the Catholic Church in Latin America and Mediterranean Europe (between state and civil society). It permits us to gauge the relative importance of each domain in given national societies. In the totalitarian regimes of Eastern Europe, for example, state power had until recently all but swallowed up the remaining domains, but a resurgent, revolutionary civil society suddenly burst from its subterranean hiding places, toppled the hated regimes and is now allowing both the corporate economy and political community to flourish in a new configuration. In many African countries, such as Kenya, a one-party state is still firmly entrenched, and a political community exists in only rudimentary form. Yet the corporate economy, though relatively weak, has its own domain of action and has built close ruling alliances with the state. Similar analyses can be carried out for each country. I shall return to the question of how this representation of the space for social practice may help us in situating an alternaitve development after completing this explanation.

Our four domains are shown to be inscribed within a bounded territory or *life space*. This is the physical space over which both the political community and the state claim sovereign power. Life spaces exist at several levels and are coincident with either political-administrative boundaries or claims to greater autonomy in governance. A stylized hierarchy of territories includes the nation, province (region), district, and city. Below the city level neighborhoods and households also claim certain territorial rights. And above the national level there may be groups of nations that form regional blocks. Territories define membership in a potential political community, and every person is simultaneously a member of several such communities whose respective "sovereignties" limit and define each other. In all cases, territory-based communities have boundaries that mark the extent of their formal authority. But these limits are always subject to redefinition: property lines may change, city limits may be enlarged or new districts incorporated, provincial boundaries may be changed by administrative ruling, national frontiers may be invaded. Finally, each territorial system defines a distribution of legal powers. Federations of autonomous territories, for example, will have a very different distribution of powers than a unitary state. There are many variations (Duchacek, 1970).

An economic space whose reach is global penetrates and overlaps with life space. It is a space structured by market relations and defined by the location of productive activities and the intersecting flows of capital, commodities, labor, and information.

Capital, which carves out the dimensions of economic space, is often in conflict with the life space of territorially bounded communities. Its spatial configurations do not necessarily coincide with territorial boundaries and systems of political and state power, that, from the perspective of capital, are regarded as *artificial*. Still, the corporate economy cannot altogether escape from territorial power and control; indeed, its continued functioning requires a legal-political system of order. In the perspective of a given territorial unit, the importance of the corporate economy to its citizens is obvious. Yet serious conflicts of interest may arise, as when a factory shuts down, disrupting the normal course of life in the city, or in cases of toxic pollution by industries unconcerned with environmental impacts in their particular localities. Territory-based communities typically seek to protect their interests, such as environmental resources, quality of life, and so forth. Because their populations are less mobile than global capital, they become saddled with the social costs of enterprise (resource exhaustion, pollution, unemployment, crime, etc.). On the other hand, global business favors an economy without boundaries; it prefers the free flow of trade, capital movements without restriction, tax holidays, generous public subsidies, and more generally a minimalist state that is broadly supportive of its claims while not overly protective of its territorial interests.

Economic space today is indeed global and is articulated through a system of "world cities," global centers of capital accumulation and control (Friedmann, 1988, chapter 3). Describing the new technologies that are helping to shape the emerging core economies of the world system around these major nodes, Peter Hall writes:

> Theoretically, such a system would permit the almost infinite dispersion of informational activities across nations and continents, including the spread of homework and telecommuting. Practically, all the empirical evidence so far points to the opposite; *technical advance will be most rapid in the existing centres for informational exchange, above all, the world cities, where demand produces a response in the form of rapid innovation and high-level service competition; conversely, peripheral regions will remain locked into a vicious circle of low demand and low innovation.* (1990, p. 18; my italics)

In other words, the dynamics of economic growth will continue to be generated in the existing major world centers – a mere 20 or so heavily urbanized regions from Paris to Tokyo – with much of the rest of the world, "remaining locked into a vicious circle of low demand and low innovation." But as we have seen in chapter 1, the world periphery comprises perhaps sixty percent of the world's population. That is the true drama to which an alternative development responds.

The four domains of social practice have traditional relationships with each other. Over the past 200 years in the West, power has been accumulating along the vertical axis a-a' in figure 2.2, linking state with corporate economy largely at the expense of power along the horizontal axis b-b', which connects civil society with political community. *The story of civil resistance told in the preceding section may thus be seen as part of the slow recovery of social and political power by households, civil associations, and social movements.* State autonomy is challenged by claims of citizen sovereignty. Odd as it may seem, the social contract theory of Locke, Montesquieu, and Rousseau, according to which the state is ultimately accountable to the people organized as a political community, is being resurrected in places where this foundation is virtually unknown, along with new claims for participatory governance.

What I have called the process of virtual exclusion from economic and political power can now be better understood as a historical process of *systematic disempowerment.* In the extreme case, disempowerment takes the form of a dictatorship backed by military power that shuts down the political community, imposes stringent controls over social movements, and tries to infiltrate and coopt civil associations of all kinds. Intimidation by the state is used to undermine collective citizen action.

Under redemocratization, some or all political restrictions are lifted, the rule of law is reinstituted, and normal civil life is resumed. At the same time, protest movements multiply, political power is reclaimed in the name of civil society, and the margins of freedom of action by state and corporations are reduced.

This drama takes place at all territorial scales. Of particular interest to an alternative development, however, at least initially, is the *local* scale, which is the privileged terrain of the disempowered sectors. Here the struggle involves a redefinition of roles between state and civil society, and civil society and corporate economy, with special attention given to new forms of political participation in planning, communal action, economic organization, and gender relations in both the household and political community.

Most important, an alternative development involves a process of social and political empowerment whose long-term objective is to rebalance the structure of power in society by making state action more accountable, strengthening the powers of civil society in the management of its own affairs, and making corporate business more socially responsible. An alternative development insists on the primacy of politics in the protection of people's interests, especially of the disempowered sectors, of women, and of future generations that are grounded in the life space of locality, region, and nation.

Although an alternative development is initially based in particular localities, its long-term aim is to transform the whole of society through political action at national and international levels. Without this quantum leap from the local to the global, alternative development remains encapsulated within a highly restrictive system of power, unable to break through to the genuine development it seeks.

Empowerment

An alternative development is centered on people and their environment rather than production and profits. And just as the paradigm in dominance approaches the question of economic growth from the perspective of the firm, which is the foundation of neoclassical economics, so an alternative development, based as it must be on the life spaces of civil society, approaches the question of an improvement in the conditions of life and livelihood from a perspective of the household.

Economic science works with a model of "economic man," understood as a rational, utility-maximizing being with a built-in moral calculus: that whatever promotes one's material interest also furthers the interest of all individuals together, provided that their actions are governed by the rules of market competition.

The starting point of an alternative development is very different. Households are composed of natural persons – that is, of three-dimensional, moral human beings who, from birth, stand in dynamic interaction with others. As moral beings all of us have certain obligations; we compete, but we also learn to work together; we relate to each other according to a complex moral code in which many of our responses are culturally patterned. As moral beings, we have not only wants or desires but also needs, among which are the psychosocial needs of affection, self-expression, and esteem that are not available as commodities but arise directly from human encounter.

Households, may be defined as a residential group of persons who live under the same roof and eat out of the same pot.[13] Each household forms a polity and economy in miniature; it is the elementary unit of civil society. Persons residing in a household may be blood-related or not. Their true families include kin who may live in households that are spatially dispersed but remain linked to each other through patterns of mutual obligation.

Each household engages in a daily process of joint decision-making. This jointness may be crystallized into a traditional gender and age division of labor, or it may be more or less open and conflictive. What is important for our purposes here is that households do collectively produce their own lives and livelihood: they are essentially productive and proactive units. This understanding of households as an economy contrasts with the model of neoclassical theory by which production typically takes place outside the home, in factory, field, or office. The household is treated as a consuming unit to which workers return in the evening and which is primarily concerned with the biological reproduction of labor and the enjoyment of leisuretime activities. Accordingly, consumption is regarded as an essentially individualizing, private activity of no particular concern to anyone, while production requires the cooperation of others and is therefore an inherently public activity subject to regulation by the state, even where the means of production are privately owned.

In an alternative development households are treated as both production-centered and public. As producing units they articulate market and nonmarket relations. As political communities they are the terrain of struggle over the allocation of household resources to particular ends, and over particular rights, such as property claims. And because as producing and proactive units they require the cooperative actions of

---

[13]   For a deeper discussion of the household concept, with citations to the relevant literature, see chapter 3.

others, community relations of households are governed by reciprocity, the most fundamental ethical principle governing social conduct (Ekeh, 1974).

In furthering their pursuit of life and livelihood, households dispose over three kinds of power: social, political, and psychological. Social power is concerned with access to certain "bases" of household production, such as information, knowledge and skills, participation in social organizations, and financial resources. When a household economy increases its access to these bases, its ability to set and attain objectives also increases. An increase in social power may therefore also be understood as an increase in a household's access to the bases of its productive wealth.

Political power concerns the access of individual household members to the process by which decisions, particularly those that affect their own future, are made.[14] Political power is thus not only the power to vote; it is as well the power of voice and of collective action. Although individuals may participate in politics on a personal basis, their voice rises not only in local assembly but also, and at times more effectively, when it merges with the many voices of larger political associations – a party, a social movement, or an interest group such as a labor or peasant syndicate.

Psychological power, finally, is best described as an individual sense of potency. Where present, it is demonstrated in self-confident behavior. Psychological empowerment is often a result of successful action in the social or political domains, though it may also result from intersubjective work. An increased sense of personal potency will have recursive, positive effects on a household's continuing struggle to increase its effective social and political power.

As its central process, an alternative development seeks the empowerment of households and their individual members in all three senses. It is therefore a process that originates both from below and within specific territory-based social formations, such as a village or barrio neighborhood. It focuses explicitly on the moral relations of individual persons and households, and it draws its values from that sphere rather than from any desire to satisfy material wants, important as these may be. An alternative development cannot be "guided" by governing elites without destroying its alternative character. It is also very different from the impersonal processes that are responsive to the principle of growth efficiency. Alternative development must be seen as a process that seeks the empowerment of households and their individual members through their involvement in socially and politically relevant actions.

---

[14] In the micropolity of the household, one can also speak of the political power of individual members, of women and men, siblings, children, and the elderly.

Giving full voice to the disempowered sectors of the population tends to follow a certain sequence. Political empowerment would seem to require a *prior* process of social empowerment through which effective participation in politics becomes possible. For instance, social empowerment, especially when oriented to women, can lead to the release from household drudgeries, and the time thus won, like any surplus resource, can now be variously applied, including to political practice. It may also contribute to an increased sense of self-confidence (as with the women who participate in the *talleres* of Santiago), which comes, in part, from conquering a fear of acting outside culturally sanctioned (patriarchal) or state-imposed norms. Ultimately, however, gains in social power must be translated into effective political power, so that the interests of households and localities can be effectively advocated, defended, and acknowledged at the macrosphere of regional, national, and even international politics.

The Politics of an Alternative Development

An alternative development does not negate the need for continued growth in a dynamic world economy. It would be absurd to attempt to substitute a people-centered for a production-centered development, or to reduce all development questions to the microstructures of household and locality. What it does do is to seek a change in the existing national strategies through a politics of inclusive democracy, appropriate economic growth, gender equality, and sustainability or inter-generational equity. In short, an alternative development incorporates a political dimension (inclusive democracy) as one of its principal ends of action. It does not make a fetish of economic growth but searches for an "appropriate" path that includes growth efficiency as one of several objectives that must be brought into harmony.[15] Appropriate economic growth sets out to optimize the use of resources over several broad and competing objectives, such as an inclusive democracy and the incorporation of the excluded sectors of the population in the wider processes of societal development. Furthermore, an appropriate economic growth path is pursued when market measures of production are supplemented with calculations of the probable social and environmental costs, or costs to third parties, that are likely to be incurred in any new investment. By

---

[15] In its most general form, efficiency is a measure of the relation of input to output. Mainstream economists assume that an aggregate measure of output, such as the GNP, is the "obvious" goal of development and that it should be achieved at the least cost, with production being measured by free-market prices. But other measures of efficiency are conceivable; for example, efficiency in relation to an employment objective or in relation to resource conservation.

differentiating territorial needs, as well as the special needs of different social groups, particularly the disempowered, it argues for a decentralized, participatory mode of decision-making in development. At the same time, it gives voice to the interests of future generations who desire historical continuity in territorial development.[16]

To a far greater extent than the prevailing mainstream model, alternative development acknowledges the *needs* and *established rights* of citizen households.[17] It regards the first as a political claim the second as a claim that has been recognized and made firm in the institutional arrangements of society. It therefore views development as not only a genuine and lasting improvement in the conditions of life and livelihood, but also as a political struggle for empowerment *of households and individuals*.[18] And even as it recognizes the importance of the microspheres of political and territorial life, acknowledging the intrinsic diversity of life, it seeks to remove the structural constraints on the possibilities for local development. This, of course, calls for a regional, national and even international politics, albeit one that has deep roots in the capillary structures of social life.

Despite its advocacy of a grass-roots politics, an alternative development requires a strong state to implement its policies. A strong state, however, is not top heavy with an arrogant and cumbersome bureaucracy; it is, rather an agile and responsive state, accountable to its citizens. It is a state that rests on the strong support of an inclusive democracy in which the powers to manage problems that are best handled locally have been devolved to local units of governance and to the people themselves, organized in their own communities.

Given the power imbalances in the world today, hope in the possibilities

---

[16] An explicit emphasis on territoriality might seem unnecessary were it not for three considerations. First, national territories are articulations of lower-order territories – provinces, districts, cities, neighborhoods – each of which, with its peculiar character, has a claim to historical identity. Second, the mainstream model of economic growth expresses the desire of global capital for an economy "without boundaries" in which there are neither organized interests nor powers that mediate between the centers of corporate decision and individual workers and consumers. In the ideology of capital, such an economy is called "free". It reduces territorial interests to a residual minimum of "law and order," such as enforcement of contracts and the maintenance of order in the streets. It also expects territorial states to deal, as best they can, with the social consequences of private investment and production decisions, such as resource exhaustion, unemployment, pauperization, pollution, deforestation, and other problems of "the commons." Third, territoriality draws our attention to the physical environment: the resource base of the economy, the aesthetic value of traditional landscapes, and the livability of the built environment in which all of our actions take place and which affects our life, directly and indirectly.

[17] Properly speaking, the language of mainstream economics lacks words for needs and rights. Its preferred and, indeed, exclusive terms are economic wants and interests.

[18] Any so-called development leads to new configurations of power. The potential restructuring of power is what impels those who expect to lose some power as a result of alternative development to resist it.

of an alternative development may seem Pollyannish. Yet if it is seen as inspiring a global movement toward a world that is more just, it may not be an impossible vision. Rich and poor countries must work together here, for rich countries have their own redundant populations, their disempowered poor. Estimates of the so-called underclass in the United States run to well over 10 percent of the total population, or to perhaps 30 million. Most are of non-European origin, and many are women heading households of their own. The politics of an alternative development is, of course, no panacea for the world's ills. In each instance it must be adjusted to the prevailing historical and cultural circumstances. But there would seem to be sufficient commonality of interests to make of its advocacy something more than a utopian dream.

3 Rethinking the Economy: The Whole-Economy Model

Alternative development means improving the conditions of life and livelihood for the excluded majority, whether on a global, national, or even regional scale. But how shall we measure such improvement? At some basic level, we can say that "improving" means simply to have more of the good things in life, though in counting them, we must be careful to subtract all the "bads" that inevitably accompany an "improvement." In a socially relevant sense, to call something "better" involves a netting out of the social and environmental costs of development. The question of who benefits and who pays is also of great significance.

At the individual or household level, "the good things in life" may be specified with some precision. We can conduct a survey, for example, to find out what would contribute to people's happiness. But at an aggregate territorial level, specification becomes problematic. People want many different things. How do we add them up? How do we make sense of this welter of information?

By convention, the most widely used aggregate measure of "the good things in life" are the national income accounts typically calculated at yearly intervals. Their power to persuade stems from their ability to reduce the immense variety of the palpable world to the single denominator of money. They accomplish this feat by using market prices to convert real goods (and bads) into the virtual good of money. National income accounts are most commonly expressed as gross domestic product (GDP) or gross national product (GNP), the difference between them being primarily in the balance of payments on foreign account. For the purpose of policy-making, income accounts may be supplemented by other data, such as employment, health, education, housing, and energy statistics. But as a comprehensive indicator of how well an economy is doing and as a measure that is closely, though imperfectly, correlated with many social indicators, it has no rival.[1]

---

[1] The literature on social indicators is vast. A more restricted bibliography on the "measurement and analysis of socio-economic development" has attempted to devise substitute measures for GNP and its derivatives. See Pipping, 1953; United Nations, 1954, 1990; McGranahan et al.,

Hence it happens that a measure of economic growth, when divided by population, has become the classic indicator of "development." Technical qualifications aside, a sustained rise in GNP per capita (adjusted for inflation) is widely understood as signifying that an improvement in the conditions of life and livelihood of the relevant population has taken place. Mainstream economists who think that they are practicing a "positive" science like to avoid language such as *good* and *better*, preferring quantitative measures that increase or decrease, gain or decline. For these economists, GNP per capita can be used as a rough measure of welfare, so long as its (usually unspecified) "imperfections" are borne in mind. Despite recurrent warnings regarding "imperfections," national income accounts have come to be used almost universally, and governments expend enormous effort in the collection of data toward their construction. Other data are collected less assiduously. In sum, there is no question that for all of their so-called imperfections, national income accounts have greatly influenced the way we think about the state of the economy, or what most would argue is "the" economy, as though national income accounts were the only accounts that mattered. Even minute variations in their fluctuating course are used as an occasion for changes in national policy.[2]

The primary message of this chapter is threefold: (1) that economic growth and development, despite conventional usage, are not equivalent concepts and should be separated analytically; (2) that national income accounts are not merely "imperfect" but actually misleading indicators, so that policy decisions based on them are more likely to aggravate than to alleviate problems; and (3) that an alternative development needs to look at economic relationships in very different ways from mainstream doctrine. If we want progress toward an alternative development, what we call the economy has to be rethought from the ground up.[3]

1972; Drewnowski, 1974; McGranahan et al., 1985; and United Nations Development Program, 1990. Some of this work is extremely interesting, and there have been a few "real-life" applications, notably in the Philippines (Mangahas, 1977). But overall social indicator work has had little practical impact.

[2]   The World Bank has been a major disseminator of this mythology, and the statistical supplement to its annual World Development Report is the most widely used source of comparative data set on national economic growth. The World Bank typically arranges countries by their rank order in per-capita GNP, allowing for a few additional categories, such as centrally planned economics ("non-reporting members") and high-income oil-exporting countries. A recent volume (World Bank, 1989a, table 30) presents some GDP estimates that, it is claimed, have been rendered more comparable by the United Nations Comparison Program (ICP), Phase V. But these supposed improvements do not address the fundamental criticisms of national economic account put forward in this chapter.

[3]   The critique of national income accounts has been going on since their inception in the 1940s, and actual accounting practices are the result of a series of compromises among contending groups of scholars. For more recent criticisms of orthodoxy, see Nordhaus and Tobin, 1972; Hueting, 1980; and Block, 1990.

1 *Measures of economic growth and development are not the same.*
Strictly speaking, national income accounts present us with data on
aggregate production growth. They tell us nothing about the meaning of
development. This argument has been frequently made from a variety
of perspectives, but we need to go over it again, because mainstream
practices are firmly entrenched, and an alternative approach needs to
be guided by different considerations.

To begin with, national income accounts don't tell us how income is
distributed either socially or territorially. As a result, we cannot know
how changes in aggregate income affect different groups of people. The
Philippine social indicators program, for example, proposes to disaggregate data according to (a) urban and rural, (b) region, (c) sex, (d) age,
and (e) family income (Mangahas, 1977). Something like this breakdown
is needed to begin to make sense of income changes in terms of the
well-being of households. We know, for example, that in most industrializing countries income growth is concentrated in only one or a few
"core" regions that dominate and exploit their respective peripheries, and
that rural areas are grossly underendowed with facilities such as for
health and education when compared to cities. So we know that national
income accounts hide from us the real and important *variations* in
income produced at meso (regional) and micro (local) levels.[4]

I want to pause here for a moment to show why territorial analysis at
meso and micro levels is in fact all important for an alternative development. Mainstream economists tend to ignore them.[5] If these economists
have a spatial conception of the economy at all, it is that of a single,
integrated (national) space in "dynamic equilibrium." Such an understanding may be a fair approximation of the interlocking core regions
of Western Europe, North America, and Japan, although even here the
periodic revolutions in production technologies upset established spatial
equilibria (Storper and Walker, 1989). But in the rest of the world most
national economies, far from having evolved integrated markets for

---

[4] National income accounts are an invention of Keynesian economists who believed that regulatory action by the state was necessary to keep capitalist economies on an even course. After
World War II Keynesianism became the backbone of central planning for the long-term development of the economy (investment planning). The national state was seen as a primary actor in
both situations. And it is a fact that, as we descend the territorial hierarchy of powers, real needs
become more apparent (because they are more amplified) even as available policy instruments
become fewer. It is also true that, if the powers of regional authorities were to be strengthened,
redistribution of income among regions would become immensely more difficult. Capital, on the
other hand, seeks the largest unrestrained markets – hence the nation or, better yet, an economic
union of nations with a common institutional framework for investment, production, and trade.
[5] Regional and urban economics have a shadowy existence as subdisciplines but have virtually
no influence on the profession's mainstream. The territorial dimension has been captured by
geography and by the hybrid discipline of regional science founded in the mid 1950s by Walter
Isard, but neither discipline ranks high in academia's pecking order.

capital, labor, information, and commodities, are typically very unevenly developed along their horizontal (or spatial) dimension.

We are interested in territoriality not because of some obscure spatial metaphysics but because people inhabit these spaces, and it is these flesh-and-blood people who suffer the booms and busts of the economy. People are not an abstract category of labor that moves mechanically at the right time and in just the right proportion to wherever economic opportunities arise. They are social, *connected* beings who live in families, households, and communities and who interact with neighbors, kinfolk, friends, and familiars. Over time, people inhabiting particular places evolve typical patterns of speech, ritual practices, and social practices with which they are comfortable and "at home." In principle, people may be mobile, as economists expect them to be. And they do move in very large numbers, just as theory predicts. But they also have local and regional ties that bind, particularly among the older age groups who have an entire lifetime at stake. They do not move easily, and when they do, it is often under duress, and the process is painful. Concerned as we are with an alternative development, therefore, we need to look closely at the prevailing conditions of life and livelihood at spatially distinctive scales.

The territoriality of development raises the further issue of what a betterment in the (aggregate) conditions of life and livelihood really means. "Better" is a qualitative judgment, and mainstream economists are afraid to pollute with their own "tastes and preferences" what they would like to believe is an objective and scientific discourse. But the opposite of scientific objectivity is not the inscrutably subjective. It can be shown that discourse can be both scientific *and* value-relevant. Indeed, economics is itself a prime example of a form of disciplinary discourse in which a whole worldview has become enshrined. And so to argue that economics, even in its neoclassical incarnation, is value-neutral is nonsense.[6]

It is granted, then, that "better" is a judgment concerning a qualitative change in the conditions of life and livelihood. But how shall we get at an answer to the question of what, in a specific instance, constitutes "better"? Far from being inscrutable, a judgment of this kind becomes rationally defensible when it conforms to three conditions: (a) it is based on full and accurate information, (b) it is based on an open and inclusive discussion among those directly concerned with the outcomes of a judg-

---

[6]   The sociologist Amitai Etzioni (1988) has made the most systematic effort to date to rescue neoclassical economics from the snares of its utilitarian premises. Following Kant, he calls his approach "deontological" – that is, informed by an ethical theory where the rightness of an action is determined *without* regard to its consequences. Deontological values contrast with utilitarian ethics.

ment, and (c) it is based on a decision-making process that is accepted as legitimate for both those participating in it and relevant others.[7]

Consequently, a necessary condition for determining the presence or absence of "development," in the specific sense we have given it here, is an *inclusive* democracy with its correlatives of civil liberties, accountability, and wide access to full, accurate, and appropriate information. It is important to point out that this political condition must prevail not only at the national level but at all lower territorial levels as well, especially in impoverished communities, where access to information cannot be taken for granted and where civil liberties may be severely restricted (Cockburn, 1977). The construction and strengthening of political community thus becomes an essential feature of an alternative development (see chapter 5).

We can now conclude this part of the argument. Development, we can say, is a more inclusive, differentiated concept than economic growth. The one cannot be reduced to the other, nor can development be encapsulated into a single, aggregate index. A diversity of criteria at different levels of territorial aggregation (and of other forms as well) is needed to determine whether development has in fact occurred (Norgaard, 1989a).

Even so, unlike economic growth, development is incapable of being objectively defined "from the outside." To arrive at a legitimate judgment of development policy (i.e., of "development" in its operational sense) requires an open and inclusive process of citizen participation. Expert judgment cannot substitute for the considered and open deliberations of citizens.[8]

2   *National income accounts are misleading as a measure of economic growth.*

National income accounts are constructed by using market prices, but market prices do not, as a rule, reflect social valuations as much as they do relations of supply and demand.[9] A striking example of how market prices can be at variance with social valuations is the interest rate. When set by the market, interest rates tend to assign a very low value to the long-term future. When capital expects to "turn around" in just a few years, the market may set the rate of real interest at 10, 20, and even more percent. But society's perspective necessarily takes the long view

[7]   Similar conditions within institutionalized science ensure that research results may be accepted as "best current posits." To an unusual degree, scientists are grouped into democratic and participatory, if mutually exclusive, communities.
[8]   This view of democratic practice has been strongly influenced by Hannah Arendt (1958).
[9]   The market prices of some commodities may of course be socially regulated through taxes, subsidies, price controls, tariffs, and so forth.

(society's "turnaround time" is generational), and the long view calls for interest rates that reflect not only the relative scarcity of capital but also the claims of future generations (see chapter 6). Projects that are uneconomic on private account because of the high rate of interest become socially feasible at interests that are significantly lower.[10]

More seriously, only costs directly attributable to capital are reflected in market prices – such as capital depreciation and research expenditures – not the costs of capital accumulation to the society at large, nor of the depreciation of society's fixed natural wealth. The costs of environmental destruction – what Roefie Hueting (1980) calls the destruction of environmental functions – are not added into national income accounts simply because there is no market for them and hence no price.

The amounts involved are not negligible. Each year vast sums are spent on repairing, or at least controlling for, the damages of an advanced industrial civilization – from resource destruction to mental breakdowns, from the spread of a drug culture, street violence, and crime to overproduction on defense technologies, competitive advertising, and excessive legal contestation (Baran and Sweezy, 1966). And poor countries are not exempt from such distortions. Yet all that society spends to deal with these and other problems that result, directly or indirectly, from unfettered and uneven economic growth are counted as gains in the domestic product. To put the matter concisely, as the real social damages of economic growth go up, so does national income. And it does so not by some accidental correlation but as a direct result of public and private action to deal with the gross malfunctioning of the system. These largely compensatory expenditures, whose object is to limit, control for, and repair the ravages of unrestrained economic growth, are typically counted as *gains* in welfare rather than costs. This is very likely why so many people perceive that we are collectively worse off as the years go by, even as GNP continues its steady climb. Their perception is not illusory: it is rather the "objective" measure of GNP that creates the optical illusion that things are getting better all the time.[11]

National income accounts also exclude vital information on production. Specifically, three types of production fail to be accurately reflected.

(a) *Subsistence activities in agriculture, forestry, fisheries, and mining.* These amounts are sometimes estimated, but estimates based on fictitious market prices ("shadow prices") and on "guesstimates" of physical pro-

---

[10]   The rate of resource exploitation is also responsive to interest rates. A high market rate will lead to a much faster rate of resource extraction than a rate that is socially determined.

[11]   One might well question why national economic accounts are distorted in this fashion. The answer comes readily to hand: to determine which costs are cost and which should be counted as a benefit would require a value or political judgment. And economic science prefers to remain on an "objective" plane.

duction are unlikely to yield even an approximate value, especially in countries where a majority of the population derives its subsistence from these activities. Occasionally, the argument is made that subsistence production is a constant proportion, rising only with increases in population, but this is hardly an empirically verified assumption. Peasant self-consumption, for example, varies a great deal with crop failures, the movement of relative prices, political stability, and economic crisis.

(b) *Informally organized production.* Informal activities cover a broad spectrum of production activities, from domestic production to micro-enterprises to a variety of illegal activities. In most poor countries they contribute to the livelihood of between 30 and 60 percent of the urban population. Diverse as they are, what informal activities have in common is their noncorporate status and their invisibility to official eyes. Even though small-sector production is basic to livelihood and even to some formal-sector production, reasonably accurate statistics on informal work are, by definition, unavailable. Production simply doesn't get reported.

(c) *Household production.* In neoclassical economics, households are seen as providing essentially two functions: they are loci of consumption and of the biological reproduction of the labor force. Unless they sell the product of their labor on the market, households are not seen as being productive. Yet the household's contribution to the national economy is in fact sizeable. For example, if all the work presently done for free in American households were to be carried out against wages, anywhere from 33 to 46 percent would have to be added to the national income (Peattie and Rein, 1983, p. 38). Most of this is work done by women and children.[12]

We cannot assume that household production for use remains constant over time. Household conditions change frequently, not only as a result of exogenous changes (as in general economic conditions) but also because of endogenous changes (the addition of new members, deaths and departures, sickness, family obligations, etc.).

For all these reasons, then, national income accounts give us an erroneous picture of production relative to the welfare of households and communities. They include as benefits what should be counted as a cost; they fail to discount declining stocks of resources; they don't reflect social valuations; and they completely disregard economic activities outside the range of the market just as they do with commodity production in the informal sector. These quantities are large and variable. Failing to include

[12] It is questionable whether attempts to come up with a "value-added" measurement of household production is a meaningful exercise, as no market exists for these activities. The figures above are cited to give an order-of-magnitude impression of the quantitative importance of housework relative to the GNP measure in common use.

them, or to include them incorrectly, yields a false conception of the economy. Large parts of people's livelihood become invisible. Work that is clearly productive, such as housework, is not seen as work at all but as a part of "consumption." There are strong grounds for arguing that the national income accounts should be revised to reflect more accurately what they currently fail to include. But there are equally powerful reasons for arguing that both the productive activities of households and the destruction of environmental functions should be kept out of the framework of national accounting; they should be monitored separately. The search for a single strategic indicator of economic health is a misguided one.

3　*Alternative development requires a "new look at the economy."*
What we think of as the economy is actually a welter of relationships that has neither a shape nor logic of its own. All economies need to be modeled before they become objectively real for us. National income accounts are such a model. That is why "rethinking the economy" is an accurate title for this chapter. It asks us to reconstitute reality in our minds. This is not to deny that very hard-edged relations exist in the business world or between landlord and peasant. An "economy" is different, however; it is a very complex set of relationships that can be modeled only by simplification. At the household level, the economy may still be reasonably transparent, but when we reach the meso and macro levels of decision-making, what we call the existing economy is very much dependent on how we choose to model it.

In the preceding paragraphs we have found that the mainstream model of the national economy is very seriously deficient from a public policy standpoint. For a number of reasons:

- It is silent on both the social and territorial distribution of income.
- It does not provide us with a measure of wealth or economic power and tells us nothing about how this power is accumulated and distributed.
- It fails accurately to reflect social valuations.
- It ignores the costs of production shouldered by third parties.
- It is silent about the costs resulting from the destruction of environmental functions.
- It excludes vital information on production arising within subsistence agriculture and other primary activities, the informal sector, and the household economy.

Thus we need to reconsider the model not by tinkering with it but by completely reconstituting it, beginning with a very different set of assumptions than the ones on which it is based. An alternative development will not come from an improved version of the model we already have but from thinking about other points of departure.

The Household Economy

To start at a point different from neoclassical economics means that we will have to find a framework that is not determined by market relations and, at the same time, bears a direct relation to people's sense of well-being. One possibility is found in the writings of Karl Polanyi (1977), for whom economic relations are deeply embedded in the matrix of social and cultural relations. According to Polanyi, economic relations "denote bearing reference to the process of satisfying material wants" (1977, p. 20). They include both economizing relations (making the best of scarce resources) and substantive relations with the environment without which human life cannot be sustained. For Polanyi, "to study human livelihood is to study the economy in this substantive sense of the term" (1977, p. 20). This methodological commitment leads him to look at institutions and, more broadly, at sociocultural relations through which our relations with the natural environment are mediated in the process of gaining a livelihood.

Polanyi's approach, which has a greatly influenced economic anthropology, leads us to the threshold of the household economy, the *oikos* from which the term economics is derived and which is also the system, central to civil society, through which nonmarket and market relations are articulated. The household economy does so by continuously solving the problem of allocating the time of its individual members to different tasks, spheres of life, and domains of social practice.

Time is households' basic resource in the production of life and livelihood, and to speak of the household economy is to see households engaged primarily in productive activities. Poor households, which are our major concern, rely heavily on nonmarket relations both for securing their livelihood and pursuing their life goals. It is otherwise with households that are better off, and whose overall economic situation allows them to place a greater emphasis on money-mediated, market relations. The extent to which households rely chiefly on money in the satisfaction of their wants is a criterion by which the somewhat uncertain dividing line – more like a broad band than a sharp edge – between the economy of subsistence and that of capital accumulation may be drawn (see figure 2.1). Here too we find an explanation of why mainstream development policy deals so harshly with the poor in poor countries. Policy is geared to capital accumulation and therefore to the improvement of the material conditions of those who are creatures of the market economy and are able to benefit from its expansion. Mainstream policy is mute on the contribution of nonmarket relations to the production of livelihood. But nonmarket relations is precisely where an alternative development has to start.

In chapter 2 I introduced a thumbnail definition of the household as a residential group of persons living under the same roof and eating out of the same pot. I referred to it as the smallest unit of civil society, as both a polity and an economy, and as a unit for making decisions on a continuing and daily basis concerning the use of household resources and other matters.[13]

Some critics argue that in their actual composition and social relations household structures are very different in different cultures, and that "the smallest unit of society" is therefore a meaningless concept. Others focus on the conflictive relations within households, fearful that by treating households as a "black box," the very different concerns and interests of men and women will be rendered opaque. Still others, thinking primarily of Northwest European and North American metropolitan settings with their high proportion of single-person households, challenge the claims of a "productive" household unit in which market and non-market relations are combined to produce a livelihood.

Despite these criticisms, I want to insist on the idea of the household as a foundation for rethinking economic relations. In what follows, I shall present a model of the household economy. Like any formal model, this is an abstraction from actually existing households and familial relations, a starting point for empirical research, not its conclusion. The first criticism is thus beside the point. Differently constituted household relations do not necessarily contradict, much less invalidate, the social production of life and livelihood which the model of the household economy purports to represent.

As regards the second argument that the household model obscures interpersonal and conflictive (but why only conflictive?) relations within households, I propose to treat households as a political economy or polity. Conflicts internal to households over questions of power – who does what kind of work, who controls what portion of whose income, whose voice should count "in the last instance" in decisions – reveal the household to be an open and permeable construct. It is, in fact, not a "box" at all but a pattern of relationships and processes that connect the household to extended family, neighbors, the market economy, and civic and political associations.

---

[13]   The household literature is extensive; the best includes H. Friedmann, 1980; Harris, 1981; Schmink, 1984; Guyer and Peters, 1987; Wilk, 1989. A less useful collection, informed by a world systems perspective, is Smith, 1984.

Finally, the argument that single-person households in advanced market economies are on the increase, comprising as much as one-third of all households in Germany and Sweden, is not especially relevant for poor people in poor countries whose lives are a long way from being fully monetized. Nor is it relevant to argue that households in the core countries of capitalism are often unstable to the point of breakdown. Households that have broken down reconstitute themselves: the single-person household is a limit situation and, in a global perspective, must be seen as an aberration. Biological reproduction and nurture typically take place in multi-person households, and productive activities yielding satisfactions but not articulated through the market continue to be an essential part of human life. It is in fact only in multi-person households that tensions between individuality and collective responsibility are acted out.

Conceptually, the "smallest unit of society" argument seeks to overcome the individualist bias of neo-classical economics with its utility-maximizing "economic man" (another abstraction!) who stands outside of all moral relations. The philosophical individualism in the Anglo-American tradition (Locke, Hume, Bentham, Mills) from which the model of "economic man" derives is far from universally accepted. A more satisfactory view of who we are as human beings and one that is consistent with the way life is actually lived in most parts of the world is one in which we are defined essentially by our social relations in ways that make it possible for us to become the individual *persons* that we are: moral beings, responsible for our actions and accountable to others. To become an individual of this second kind − a person and not an abstract utility-maximizing creature − membership in a household and in the wider relations of family and clan are of central and lasting importance. Making households the starting point in the production of life and livelihood is to emphasize this relational view of ourselves.

Households are miniature political economies that have a territorial base (life space) and are engaged in the production of their own life and livelihood. Households are *political* because their members arrive at decisions affecting the household as a whole and themselves individually in ways that involve negotiating relations of power. They have a *territorial base* because people have to have a place to live even if it is only a cardboard shack or a bit of pavement under the open sky. Their life space may only be temporary and neither very secure nor properly their own, and it may be shared with others. But every household activity requires such a space, and households that have continuity in time also claim and defend a life space of their own. Finally, households are conceived as *producers* and thus as a collective actor on behalf of their own (and sometimes others') material interests.

To take this conceptualization further, households are composed of *individuals* who stand in relation to others in ways that are defined by the principle of reciprocity and mutual obligation. These others may be kin, quasi kin, neighbors, and friends, but they are also found among the members of local organizations with whom we interact: schoolmates, workmates, clubs, the soccer team, religious congregations, and so forth. It is because our lives are, in fact, only partially defined by market relations and are, outside the market, governed by moral rules that are learned, first in the intimacy of the household but later in other settings as well, that James C. Scott (1976) can speak of a "moral economy" of mutual exchange.

To stress the moral economy is not to overlook household relations with markets. A substantial share of household resources must be allocated to produce in and for the market economy. In no sense, then, should we regard households as somehow set apart from markets. In the economy of the household, moral and market relations are closely articulated with each other.

In speaking of a household's allocation of the time, skills, and income of its members, we must bear in mind the interactive domains of social practice introduced in chapter 2. In addition to markets and the moral relations of civil society, households may allocate the resources at their disposal to both the state and the political community. I will now consider household relations to markets and civil society (figure 3.1 shows a model of these relations). Following Canada's Vanier Institute of the Family, which first introduced the concept (1983), I call it a model of the *whole* economy.[14] It is a look at economic relations from a household perspective.

The Whole-Economy Model

Two economies are shown to intersect each other: the economy of capital accumulation and that of subsistence (see also figure 2.1). The area of overlap is the informal economy, which may be directly linked to capital accumulation through the formal production sector (subcontracting) or may remain largely peripheral to it.

Households are shown with three major sources of monetary income: formal work, informal work, and net family transfers. To arrive at disposable income, taxes (if any) must be deducted from the total (although some taxes may be returned to households via subsidies and other transfers), along with capital expenditures on informal production

---

[14]   The form of the model is an adaptation and elaboration of Ignacy Sachs's proposal (1988) for defining what he chose to call the "real" economy.

where appropriate. A more complicated accounting problem arises from households' moral obligation to make income transfers to nonhousehold family. Through open-ended, "univocal" reciprocity, these transfers may eventually be paid back in some fashion, but, as a rule, no precise accounting is rendered.

The box on the right-hand side of figure 3.1 shows the variety of inputs into the production of a household's livelihood. Obviously, money is required here, but so are in-kind contributions from domestic and communal work. Three kinds of household expenditure are distinguished: (a) for consumption proper (such as food and most clothing); (b) for investment in household durables, including housing, furnishings, and equipment (DG + H); and (c) for investment in the capacities and skills of individual household members (human resources, HR). Both forms of investment are typically funded out of current household income and, in addition, may require physical labor (sewing school uniforms, building housing). Expenditures on them are here treated as investments because of the very long time period, up to two generations, over which returns on the investment are fully realized.

The "production-of-livelihood" box also shows time inputs to domestic work (D) and the communal economy (CE). The former acknowledges the importance of domestic labor in such activities as shopping, cooking, nursing, collecting water and firewood, cleaning, raising infants and small children, making repairs, construction, vegetable gardening, and similar tasks done in and around the household. The latter involves participation in productive activities within the barrio or village, such as preparations for community celebrations, the construction of schools and playgrounds, the improvement of access roads, and the collective preparation and distribution of hot meals (*ollas comunes*).

Finally, the box shows possible transfers from the state, such as health services, land donations, subsidized bus transportation, school lunch programs, and police protection, all of which are usually made in kind.

Using this model we can trace changes in the decisions of households, especially poor ones, when households respond to changes in both external and internal conditions affecting their lives: a prolonged economic crisis, for example, or births, deaths, illness, and departures.

- Women may enter the labor market at lower pay than men, or start to engage in informally organized activities (perhaps by taking in washing or working as a seamstress), thus reducing available time for domestic work in the household (D).
- Expenditures for household durables, housing construction and maintenance, as well as human resource development may be scaled

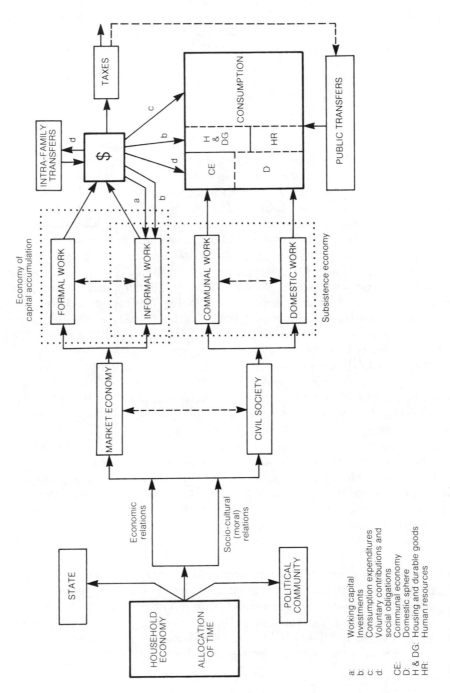

a: Working capital
b: Investments
c: Consumption expenditures
d: Voluntary contributions and
        social obligations
CE: Communal economy
D: Domestic sphere
H & DG: Housing and durable goods
HR: Human resources

Figure 3.1   The whole-economy model

back, reducing households' long-term chances for success, while investing in informal activities, both legal and illegal, becomes central to household survival.

- Per-capita household consumption may be reduced, although reductions may not be equally spread among household members. Women, especially young girls and children generally, may be the last to eat. Abortions (infanticide?) may become more frequent.
- Contributions to the communal economy, especially in work effort, may increase as households seek to lower their consumption costs by participating in community enterprises (community gardens, day care, *ollas comunes*, etc.) that help households meet their most urgent needs.
- In the case of crop failures, civil war, and other disasters, rural people may migrate to cities, either singly or in households, for food and safety as well as in the hope of finding work that will allow them to send money to any kin remaining in the countryside.

Reverse tendencies may be observed when the formal economy picks up again, agricultural conditions improve, and growth is resumed. As the emphasis in household decisions moves from subsistence to accumulation, household investment in durables and human resources increases. Informal activities that are marginal to the economy are abandoned, while those which are linked into the formal economy through subcontracting are likely to prosper. Enthusiasm for community enterprise also declines, allowing more time to be spent on domestic work. Still, some community obligations continue to be honored, because trust and reciprocity need to be established in case another crisis occurs and also for reasons of genuine neighborly affection. Solidarity, especially among women, often transcends a purely utilitarian calculus and may be maintained simply for the pleasures it affords: this kind of relationship is psychologically empowering. At the same time, some women, having gone into the economy, may not be keen to return to domestic drudgery and continue to work, albeit at very low wages, outside the home, partly because it gives them greater financial independence and partly because their workmates' company has become important to them.

Lessons for an Alternative Development

This sketch of the whole-economy model has focused on households' production of livelihood and, in a broader sense, of life itself, because economic activities are not abstracted but merged with other life-generating forces. How does this help us gain a clearer understanding of alternative development? To answer this question, let us review some of the model's salient characteristics:

- Central to the production of people's livelihood are households rather than the abstract, individual carriers of labor power that mainstream economic theory often refers to as "factors."
- Households' activities are not geared to limitless accumulation, but to the support, maintenance, and betterment of its life and livelihood, usually turning on very specific, concrete objectives only some of which are capable of being traded for money.
- Consumption is not an activity that can be separated from production, as in neoclassical economics, but is merged with, and is therefore largely indistinguishable from other household activities.
- As producers of their own life and livelihood, households are viewed as proactive. In neoclassical theory, by contrast, their principal task consists of (passive) consumption and the biological reproduction of labor power.
- Households are connected to at least some other households through relations of trust, reciprocity, and social obligation. First in line are relations with kin and friends but also with nearby neighbors, especially those whose economic situation is roughly comparable to one's own. Second in line are functional solidarities with companions in the labor movement, political organizations, and other nonterritorial associations.
- The household is both localized and connected to wider spheres of practice. In the *social domain* linkages may be made through religious and ethnic and territorial organizations. In the *economic domain* they are made through participation in syndicates, small-business organizations, and in labor markets. In the *political domain* they are made through social movements and political organizations. Life-space concerns are thus acted out on stages larger than those of village and neighborhood. Although it is correct to say that the further activities move from the matrix of the household economy the more segmented they tend to become, they can be correctly understood only if they are traced back to their origins in the political economy of the household.
- Household relations dialectically articulate a series of opposites: rational (economic) with sociocultural (moral) relations, formal with informal activities, life space with economic space, and the economy of capital accumulation with the economy of subsistence.
- A central problem for households is how to allocate scarce resources of time, skill, and money among members and the four domains of social practice. Time allocation is central to the functioning of the

household economy, and traditional allocations give rise to persisting gender and age labor divisions. In time, however, these patterns may give way, because household economies are also political communities, where in central questions such as time allocation must be continually (re)negotiated.

- A major role in the household economy is assigned to women who, normally dominant in the household itself, are also increasingly visible in the wider spheres of community life and in informally organized work.

This model of the whole economy is directly relevant to an alternative development. It defines a point of departure outside the mainstream of economics:

- alternative development recognizes the interdependencies which exist between the rationality of economic reasoning and the moral relations that tie people's fates to each other at the intimate scale of kinship, friendship, and neighborly community;
- alternative development signals the importance of people's life spaces in the production of their livelihood as well as their deep interest in bettering their conditions of life, including the quality of their immediate physical environment;
- alternative development emphasizes the need to articulate the social valuations reflective of each community's life world with the abstract valuations of the market;
- alternative development acknowledges the need to honor the claims of future generations in present decisions as a legitimate expression of households' desire for inter-generational continuity.

We may think of alternative development as a series of *loops of articulation* at different scales – loops that spiral outward from the household economy to progressively larger domains of social practice (including the economy dominated by corporate power at macro and meso levels) before returning to the microeconomy of local communities where households and their immediate concerns become again visible. These loops of articulation seek to strike new balances between dialectical opposites. They are also ways by which the interests, concerns, and values arising from within the microsphere of households, with their legitimate political claims for economic and political inclusion, penetrate the meso- and macrospheres of social practice and thus also the spheres of established relations of power. This penetration is bound to be conflictive.

In the next chapter, I will take this model of the whole economy further to a consideration of the meaning of poverty and how, in alternative development, we should be thinking about people who are called poor, and what actions should follow from such a rethinking.

The argument is that households, not individuals, are "poor," and that poverty itself can be redefined as a state of disempowerment. The question of empowerment is then discussed in terms of household access to the bases of social power, and the implications of this reinterpretation are traced for an alternative development and the role of the state.

4 Rethinking Poverty: The (Dis)Empowerment Model

Without the prevalence of real poverty in the world, there would be no need for an alternative development. Reforming the existing system would be sufficient to reduce vestigial injustices, to take care of social needs identified in public discourse, and to reduce environmental degradation by carefully dosed state interventions in markets. But massive poverty unquestionably exists as a worldwide issue, and so the issues of an alternative development must be addressed.

The trouble is, we think that we know about poverty, and that all that remains is to think up better ways to do ... what? *Eradicate it? Reduce it? Alleviate it? Cope with it? Manage it?* Quite aside from being unsure what it is that we want to do about poverty, we are wrong to think that combatting poverty simply boils down to *knowing how* without, at the same time, being very clear about the *what* of poverty. We need to know what causes poverty, and whether poverty is one big or many small questions. There are urban and rural poor, and for a few, poverty may be a chosen way of life. Others, though they might live poorly by some standards, don't think of themselves as poor. Still others are poor temporarily, while many who are born into poverty do not expect ever to escape their condition and have come to accept it as in some sense a natural condition. Nor is poverty viewed everywhere, as it is in Western countries, as a radical evil that prevents the poor from "human flourishing."

These are some of the issues that need to be separated and clarified. What we are looking for is an understanding of poverty as a public issue that must be approached collectively.

Bureaucratic Poverty

Poverty is traditionally defined by those who regard themselves as the social superiors of the poor. Hence it comes about that the poor are widely regarded with suspicion. The nineteenth century called them the

dangerous classes, haborers of indolence and vice. These negative perceptions have not disappeared. To be poor is still widely perceived to be
dirty, dumb, dirty, wanting in skills, drugged, prone to violence and
criminality, and generally irresponsible. People with these traits must be
controlled, institutionalized, and managed. They fill our jails, and when
the jails are full, more jails are built. In the United States, where much
existing poverty is also concentrated among certain ethnic groups, some
of the poor are defined as members of an "underclass" who have few if any
rights and are perceived as an immanent threat to the established order.
At least one well-known commentator wanted to put them all into
detention camps (Banfield, 1970).[1]

Social reformers have protested the identification of the poor with vice.
They argued that the poor are not themselves to blame for their condition, which should instead be regarded as a result of unfortunate
circumstances. In the reformers' rhetoric, the dangerous turned into the
unfortunate, or disadvantaged classes. Because abject poverty is degrad-
ing, they argued, poor people should be raised at least to the level of a
decent minimum, with a roof over their heads, clothes to cover their
nakedness, and food on the table. Other reformers concerned themselves
with issues of work and employment. The poor were poor, they said,
because there weren't enough jobs of the right kind. And so, not un-
reasonably, these reformers concentrated on policies for full employment
and on "work and welfare" programs.

But, in the final instance, the guardian of the poor, at least in modern
times, has always been the state, which, on behalf of the rest of society, is
charged with managing the poor, controlling the dangerous classes, running the jails, and administering the welfare program under its authority.
Labor has becomed divided, so that the poor are now the primary
responsibility of the police and an extensive bureaucracy of social workers, parole officers, and other specialists. Ordinary citizens are thus
relieved of their responsibility for the poor, confident that the problem is
being handled by qualified professionals.

The state bureaucracy has evolved its own language to describe its
relationship to the poor. Some of these terms have gained wide currency
and now form part of the standard vocabulary of poverty.

● *Poverty line.* The level of minimum household consumption that is
  socially acceptable. It is usually calculated in terms of an income of
  which roughly two-thirds would be spent on a "food basket" calcu-

---

[1] The term *underclass* has come into common usage in the United States, were major private
foundations are sponsoring research on this heading. The term itself has never been properly
defined. In a general way, it stands for the hard-core poor, most of whom belong to disadvantaged ethnic groups. The underclass is America's locked-out minority.

lated by welfare statisticians as the least-cost provision of essential calories and proteins. In setting this standard, however, statisticians do not count the labor time required for either least-cost food acquisition or the preparation of meals. Thus they tacitly assume the presence of women who are willing to undertake this work without pay, or women's subsidy. They are usually unwilling to acknowledge that time allocated to these activities could be used to earn additional household income.[2]

- *Absolute and relative poverty.* To fall below some minimum standard of consumption is to be regarded as absolutely poor, or indigent, and thus dependent on charity. But poverty also exists *above* this line, although that kind of poverty tends to be regarded differently, being judged in terms of the distance between the poor and nonpoor or of relative income. Other descriptive categories used include the *chronic poor* and the *borderline poor.*

- *(Un)deserving poor.* This terms carries over the traditional European view of the poor as lazy and improvident, inclined toward vice. The "deserving" poor are those who are willing to conform to the expectations of the nonpoor, are clean and honest and responsible, willing to accept any kind of work for any wage offered. The deserving poor are pictured as docile and industrious. They alone, it is argued, have a right to charity that will help them "to stand on their own feet."

- *Pockets of poverty.* One of the beliefs informing the work of welfare bureaucrats is that the poor are comparatively few in number and that the problem is therefore manageable. The phrase "pockets of poverty" was invented to suggest that the problem is relatively insignificant. "Pockets", it is suggested, can be "mopped up." Mass poverty is another matter altogether. It is a term the bureaucracy would rather avoid.

- *Target population.* This refers to a specific group of people that is made the object of government policies and programs. They may be female-headed households, children, landless rural laborers, small-farm peasants, victims of war or drought, and shantytown dwellers.

Those labeled as the poor have few options other than to acquiesce in the role assigned to them as the state's wards. For the most part, they are regarded as incapable of taking charge of their own lives. As in so many

---

[2] In some countries this problem is avoided by simply calculating the poverty line in terms of absolute consumption, without income conversion. According to Rodgers (1989), measures of absolute consumption are "both easiest to use [as an indicator of poverty] and most widespread; and while [they] can be readily criticized, simplicity and data availability are virtues not to be despised" (p. 5).

cases of service professionalism, welfare clients are treated as helplessly dependent on the authorities who have set themselves up as their caretakers. It is their clients who sustain and provide the raison d'être of a large and inevitably growing administrative apparatus that, although responsible *for* them, is never accountable *to* them.

The attention of international aid professionals is, for the most part, preempted by the "absolutely" poor. What this means, in practice, has been explained by M. S. Ahluwalia of the World Bank (Ahluwalia, 1974) at a time when that organization was beginning to show some interest in poverty issues:

> The incidence of poverty in underdeveloped countries defined in absolute terms has powerful appeal for dramatizing the need for policy action in both domestic and international spheres. Estimates for this type have been attempted for some countries using arbitrary poverty lines for each country to measure population below these levels.... For each country we have estimated the population living below ... [annual] per capita incomes of U.S. $50 and U.S. $75 (in 1971 prices)...the countries included ... account for about 60 percent of the total population of developing countries excluding China. About a third of this population falls below the poverty line defined by U.S. $50 per capita *and about half falls below U.S. $75 per capita.* (1974, pp. 10–11; my italics)

By World Bank reckoning, therefore, the difference of a mere $25 per capita more or less – the price of a good meal in Los Angeles – meant that between 15 and 20 percent of the population was either counted or not among the "absolutely" poor; the difference amounted to more than 200 million people.[3]

Given the overwhelming magnitude of poverty in poor countries, 50 percent according to World Bank estimates in the 1970s and 60 percent according to our calculations for the 1990s, it becomes conceptually difficult to isolate the poor. In the poor countries of the world there are no administratively convenient "pockets of poverty." The poor form a majority. They are the peasants and popular urban sectors. They are the people.

The World Bank did not draw this conclusion, however. To do so would have led it to abandon its traditional approach to economic growth and "development." And this, even under Robert S. McNamara's relatively enlightened reign from 1968 to 1981, it was not prepared to

---

[3] Had the World Bank applied the equally arbitrary standard of $100 per capita, perhaps two-thirds of the population in the world's poor countries would have been "absolutely" poor.

do.[4] Instead it adopted the perhaps not unreasonable idea that a redistribution of income toward the poor might be easier to carry out if there were rapid economic growth. But, of course, the World Bank also noted, as economist Simon Kuznets had shown, that, in the initial phases of growth, income inequalities might actually widen. Operating within the mainstream of international economic assistance, therefore, it proposed a triple strategy of *accelerating economic growth, reducing population growth,* and *redistributing income toward the poorest sectors of the population.* To the extent that it relied on redistribution, the bank aimed at increasing the consumption levels of the poor.[5]

Poverty and Basic Needs

In the mid 1970s basic-needs approaches became the axis around which virtually all proposals for an alternative development turned. The question was put on the agenda by the International Labour Office (ILO) when it convened a global conference on employment, growth, and basic needs in 1976 (ILO, 1976a, 1976b, 1977). At the time ILO efforts were paralleled by the World Bank's own strategy of "redistribution with growth" (McNamara, 1973; Streeten and Burki, 1978; Ayres, 1983). It was a brief period of efflorescence for an alternative development. By the end of the decade, however, the winds were blowing from another direction.

The basic-needs approach has been subject to multiple interpretations. The official view presented at the 1976 conference defined basic needs to include

- minimum requirements of a family for private consumption (food, shelter, clothing, etc.);
- essential services of collective consumption provided by and for the community at large (safe drinking water, sanitation, electricity, public transport, and health and educational facilities);
- the participation of the people in making the decisions that affect them;

[4] Writing of the McNamara years, Robert L. Ayres thought that the underlying rationale for the World Bank's poverty projects were "political stability through defensive modernization. Political stability was seen primarily as an outcome of giving people a stake, however minimal, in the system. Defensive modernization aims at forestalling or preempting social and political pressures. If defensive modernization is successful, it results in conservatism among the newly modernized and thus to [*sic*] their contributions to political stability" (1983, p. 226).

[5] To move the bulk of the poor from $50 to $75 per capita might require several decades of sustained economic growth. But what actually happened during the 1980s was a *decline* in the real incomes of the poor.

- the satisfaction of an absolute level of basic needs within a broader framework of basic human rights; and
- employment as both a means and an end in a basic-needs strategy. (Ghai, 1977)

This, the ILO said, was a minimum definition. Even so, the public presentation of this vision stirred up a good deal of controversy. Some representatives of industrialized market economies and employers' delegates thought that the ILO was overemphasing structural change and redistribution as essential requirements to meet basic human needs. They considered rapid economic growth as the most important remedy instead (ILO, 1977). Others saw in basic needs the key to a conception of alternative development that linked into the Cocoyoc meeting, the Dag Hammarskjöld Foundation's report *What Now?*, and similar contemporary statements.

The debate over basic needs was highly conscious of the political implications of the proposed approach. National elites would not look kindly on efforts to reduce their share of the national product in favor of social classes that had been excluded from development (Bell, 1974). A widely shared view held that elite acquiescence in a basic-needs approach would be a great deal easier to achieve if rapid economic growth were not itself being threatened.[6]

In the heat of this debate, the meaning of basic needs in more conceptual terms tended to get lost. But for an alternative development, more than code words are needed. If basic needs are relevant as a planning concept – and we have not yet established this – what should we take them to mean? The history of the concept takes us back to the first social survey ever, undertaken more than a century ago, in 1886, when Charles Booth presented his typology of the urban poor in London as a basis for designing public policies tailored to their condition (Hall, 1988, pp. 28–31). Booth's work initiated a long line of research into levels and standards of living, worked its way into the construction of poverty lines, was central to a revival of welfare economics in the post–World War II era, and finally issued in what came to be known as the social indicators movement of the 1960s and 1970s, which proposed a greatly expanded program of statistical research into social conditions at both national and metropolitan levels (Gross, 1966; Perloff, 1985; United Nations, 1990).

The basic-needs concept was an outgrowth of this earlier research into levels and standards of living. It may be useful at this point to introduce the concept in a more formal way, for what we call "basic human needs" can have several meanings, each with different policy implications. Only

---

[6]   No one, of course, challenged the continued dominance of existing elites.

one of these, however, was seriously considered as the basis for a new approach to development. I shall begin with the concept of *need*, a word that can be used in four different senses:[7]

1   *As an intense want* (needs-1). The social process through which needs-1 are identified is the market and related market research. The unit of "wanting" here is the individual in an actual or potential market transaction. Individuals may or may not have the means to satisfy their deeply felt wants, and when they are unable to do so they are frustrated or disappointed.

2   *As a functional relationship* (needs-2). We say that A is needed to accomplish B, as an appropriate tool is needed for hammering in a nail. The relationship is that of an appropriate means to an end. Needs-2 are typically identified by experts who base their decisions on either scientific-technical principles or experience. Thus when nutritional needs-2 fail to be met, degenerative diseases are likely to follow. Or when educational needs-2 are not met, a person's chances for employment and income may be reduced. In general, failure to meet functional needs-2 leads to an impairment in human/social performance characteristics relative to the end in view.

3   *As a political claim* (needs-3). This is a claim made by a group on resources that are managed in the common interest of a political community. Needs-3 (e.g., agricultural subsidies, research on AIDS, military weaponry, afforestation) are thus turned into a political argument. The political claim is for a (re)allocation of the collective resources in favor of the claimants. Political claims can be argued in terms of needs-2, but a more directly political argument may also be used, including needs-4.[8]

---

[7]   For the philosophical basis of the needs concepts, see Braybrooke, 1987; Leiss, 1976; Heller, 1976; and Soper, 1981.

[8]   Like all politics, the politics of basic needs requires a political community that sets up certain rules of governance and then defines who is to be included as a citizen member. The nation is the best known of political communities, but political communities also exist at household, city, district, region, and even supranational levels. As citizens we claim reciprocal entitlements to common resources. In return for our obligations to the community (to pay taxes, to render service, to defend the community against aggressors), we may press for certain rights or entitlements, such as free public education or subsidized transportation services for the elderly. The first, already enshrined as a right in virtually all countries, may be effectively denied to some, who find its financial demands unreasonable. To attend public school, for instance, children may be required to buy expensive uniforms and purchase textbooks, which would effectively (and inequitably) exclude many poor children from school. The second claim, for subsidized transportation, presents a different sort of argument, for seniors citizens would be competing with the claims of other groups within civil society. Because "rights" have not yet become established in the transport area, seniors would have to present themselves as, for example, agrieved citizens who have served their community well and now wish to enjoy some privileges in return.

4   *As a customary right* (needs-4). Successful struggles for needs-1 and -2 may result in securing broadly based rights or entitlements, such as free public education, health services, clean drinking water, public transportation, social security, and so forth. Needs-4 are claims that have been politically accepted and institutionalized. They set up patterns of expectation among the population that, if broken, result in popular outrage over what is perceived to be an unjust infringement of their rights.

Major contenders for a basic-needs approach are clearly needs-2 and needs-3. The former are identified by professionals, the latter by people acting through the political process.[9]

Given the origins of the basic-needs approach to development within the international bureaucracy, it is not surprising that it was needs-2 that became the principal focus of attention. Needs in the sense of political claims was not a concept bureaucrats knew or with which they cared to work at least in relation to poverty.[10] The world's poor would thus continue to be patronized by the rich. Priorities would be set for them from outside their own communities. Aid levels (and forms of aid) would be determined unilaterally.[11]

When the welfare bureaucracy is asked to define what is basic about basic needs, it is in effect asked to set priorities for investment, production, services, and consumption. But how does it go about this task? One thing is clear from the start: basic needs must be affordable. The means for satisfying them must already exist, or they must be created. Poor countries are unable to afford what might be considered basic needs in countries that are rich. To have running water and electricity in every

---

Needs-3 are clearly not as "objective" as are functional needs. The grounds of argument are often subtle. Civic discourse objectifies needs-3 by requiring each claimant to present the best possible arguments in public. Needs-3 are therefore more than simply a reciprocal entitlement. They must be decided by a process that gives equal voice to all members of a polity, so that both obligations and rights may be decided through an open democratic process. According to Soper (1981), "all the definitions of [needs] and all appeals to the concept of 'human needs' are, and must necessarily remain, problematic.... [I]t is only when the question of human needs is posed in a form that recognizes this problematicity, that is posed in its full dimensions, and that means, importantly, in the acknowledgment of its *political* dimension" (p. 1; my italics).

[9]   Lee (1977) additionally identifies needs-1 as relevant for a basic-needs approach. Here, needs would continue to be set by experts (as with needs-2), but experts would have to pay attention to consumer tastes and preference, so that the market would clear any production geared to the satisfaction of needs-1.

[10]   To my knowledge, the only presentation of a case for needs-3 is an essay of my own (Friedmann, 1979). For a related argument see Soper (1981).

[11]   International aid is, of course, never determined unilaterally in the strict sense, but the counterparts of international bureaucrats are principally national bureaucrats; the people themselves are never consulted.

home is not a currently affordable option in Mozambique or Bangladesh. And there are other differences as well. People living in the tropics do not have to make provision for space heating; those living in temperate climates cannot survive their winters without it. In addition, tastes and preferences vary a great deal and must be included in the equation.

These considerations would seem obvious, and in the 1970s planners readily acknowledged that the responsibility to define basic needs and to set priorities for their attainment would have to be delegated to individual countries (Streeten and Burki, 1978, p. 413). They did not, however, press this logic of devolution to groups below the nation level. If planning for basic needs is to be done separately for each country so that means may be adapted to ends and ends to means, both reflecting relative urgencies and scarcities, why should the same not be true for the country's regions, cities, towns, and villages?[12] Basic-needs planning is not simply a matter of declaring that, say, an average of 2,200 calories a day is necessary for a healthy and active life. An abstract datum such as this needs to be translated into a national food security policy and a program for getting enough of the right kinds of food to *every* household in the country. And to do this properly requires a territorially differentiated approach that goes beyond even food security to basic questions concerning development: the relative importance of markets; interregional and local transport systems; agricultural organization; the relative importance of investment in food production as exports; territorial self-reliance; and so on. These matters need to be considered in excruciating detail at all levels of territorial governance and planning and will ultimately have to enlist the will and energies of the people, without whose active collaboration nothing of lasting value can be accomplished. This, roughly, is what is called multilevel planning in India, except that the vertical integration of planning in such a system cannot be taken for granted, but involves a continuous *political* process in which different territorial, no less than sectoral, interests contend. In all this it is far from certain that the interests of the absolutely poor – who are typically excluded from both politics and planning – will be given the kind of attention implied by the basic-needs approach.[13]

[12]   To mention self-governance in this context is, of course, questionable. Although virtually all countries are capable of planning at regional levels, formal planning abilities tend to deteriorate quickly after that. All the same, the logic of devolution in setting economic priorities is not invalidated by the existing lack of appropriate institutional arrangements for local governance.
[13]   International agencies struggled mightily to define hierarchies of basic needs. Streeten and Burki (1978), for example, distinguish between what they call core needs (food, water, clothing, shelter) and all other needs. One supposes that the Streeten-Burki definition of core needs is influenced in part by the ability of international organizations to influence their provision – for example, urban housing for the poor – rather than by more philosophical considerations, where meeting food requirements may have to be weighed in a comparison with, perhaps, liberty.

If nothing else, the intense debate over basic needs revealed the dilemmas facing development planners in the mid-seventies. These dilemmas were resolved, albeit temporarily, reflecting the current balance of power among contending interests. Among the more important conflicts were these:

- *Economic growth vs. (re)distribution.* To what extent does rapid economic growth, with its implied structural transformation of the national economy, require a markedly unequal distribution of income and wealth? The issue was "resolved" by the unleashing of market forces under the banner of neoliberalism in the 1980s. The redistributive role of the state was minimized, while income inequalities, along with unemployment and landlessness, increased and real wages declined precipitously.

- *Technocratic vs. political determination of basic needs.* Should basic needs be defined along functional lines by experts and planners, or should they be identified through open discourse by each territory-based community for itself? Basic needs must be worked into the allocation of common resources. Is allocative planning simply the outcome of a process of competitive "claiming" by organized groups of citizens within the limits set by a democratic politics, or of technical calculations? This issue was unambiguously resolved in favor of technocracy. The basic-needs discourse arose from within the international bureaucracy. It was unthinkable that this bureaucracy should offer to subordinate its work to a democratic politics that at the international level did not even exist. National planning would be done in imitation of international practice. The hoped-for revolution of basic needs was to be a revolution from the top.

- *Production vs. consumption.* Should planning for basic needs be oriented primarily to increasing individual household consumption, or should it be seen as directing resources toward improving the productive capacities of the poor in informal urban activities and small-scale peasant farming? This debate assumed that consumption could be split from production as a meaningful activity in its own right, just as it is in the national economic accounts. The resolution tended to favor production. At the insistence of the IMF, whose role in shoring up poor, unstable economies was greatly enlarged with the debt crisis of the eighties, many countries were obliged to eliminate consumer subsidies for food, urban transportation, and so forth, while social programs, especially for low-cost housing, were curtailed. Enthusiasm waxed over the rediscovery of informally organized work as an enterprising, indigenous form of petty-capitalism. Small-farmer strategies were also encouraged.

- *Markets vs. planned allocation.* The basic-needs approach would have required extensive government planning to favor those population sectors whose ability to participate in markets was extremely weak (we may recall that the World Bank had estimated that 50 percent of the population in poor countries had annual incomes of less than $75 per capita). Or, as Streeten and Burki (1978) put it, "The emphasis of basic needs on restructuring production, not necessarily in response to the preferences of the people with very unequal incomes in an imperfect market, implies a substantial role for government" (p. 414). But in the wake of economic crisis, planning by the state was becoming discredited, and the resolution favored markets. Thus it became fashionable to say that the state was part of the problem and to stress its corruption, incompetence, and political instability. Publicists like Peruvian Hernando de Soto argued that his country was still caught in the vise of a precapitalist economy organized along mercantilist lines, in which powerful domestic interests were protected by the state. What Peru needed was more of the free-wheeling, enterprising spirit so valiantly displayed by the tens of thousands of informal entrepreneurs on Lima's streets (De Soto, 1989). International aid agencies echoed this passionate call for unfettered market competition in an open economy. The protective, redistributive role of the state had to be curtailed.[14]

As a result of this series of "resolutions" favoring accelerated economic growth, technocratic decision-making, production, and markets, the basic-needs approach has become virtually inoperative. References continue to be made to the "unmet needs" of the world's poor (World Commission, 1987), but there is little action on this front.[15] And so the condition of the poor continues to deteriorate (Rodgers, 1989).

[14] De Soto's (1989) argument was especially appealing to free marketeers. It seemed to offer a way out of the otherwise embarrassing dilemma that the liberalization of the economy, combined with export-led growth, would further isolate the poor on their "reservations" – squatter communities on the outskirts of large cities – even as it would generate increasing numbers of landless laborers in a rapidly modernizing countryside. De Soto's argument, that the poor in the city's informal sector, far from leading unproductive lives, display enormous ingenuity and talent in generating a livelihood for themselves, was an appealing message in the ideological struggle. The poor could now be portrayed as the vanguard of a yet-to-arrive entrepreneurial capitalism, not only in Peru but in the rest of Latin America as well. Indeed, the subtitle of his book made an even more sweeping claim: "The Invisible Revolution in the Third World." De Soto was thus heralded as a major player by, for example, San Francisco's International Center for Economic Growth, which supported the publication of his book in English. The objective was to deconstruct the Keynesian state and to give rein to the free market. De Soto's book made it possible to put up a smokescreen around the devastating effects of such a policy on the poor.

[15] Even the environmental movement, of which the Brundtland Report is an expression, is divided within itself. Deep ecologists, for instance, have an Earth First! attitude in which the real problem of humanity is humanity itself.

But the debates of the seventies were not all in vain. They have left us with some firm conclusions about poverty that can lead us toward a major rethinking of the question. We have learned, among other things, that

- basic needs are essentially political claims for entitlements;
- growth-maximizing strategies are not in themselves sufficient to satisfy these claims, even though rapid growth, as in the Republic of Korea, Taiwan, and Singapore, is compatible with relatively low indices of income inequality;
- poverty is a multidimensional phenomenon and does not signify merely a relative lack of income;
- greatly improved statistical systems are needed to assess people's quality of life and to contribute toward defining appropriate standards of living;
- the poor must take part in the provisioning of their own needs rather than rely on the state to solve their problems;
- to become more self-reliant in the provisioning of their own needs, the poor must first acquire the means to do so; and
- effective antipoverty programs cannot be devised at the top for implementation downward through a compliant bureaucracy but must emerge from the hurly-burly of politics in which the poor continuously press for the support, at the macro level, of their own initiatives.

All these lessons have contributed to a new perception of what it means to be poor. From the perspective of alternative development, the poor are no longer regarded as wards of the state but as people who, despite enormous constraints, are actively engaged in the production of their own lives and livelihood.

Poverty as (Dis)Empowerment

The (dis)empowerment model of poverty is a political variant of the basic-needs approach. It is centered on politics rather than planning as the principal process by which needs are identified and the means for their satisfaction pursued.

The starting point of the model is the assumption that poor households lack the social power to improve the condition of their members' lives. It places the household economy into the center of a field of social power in which its relative access to the bases of social power may be measured and compared (figure 4.1). These critical terms require further explanation.

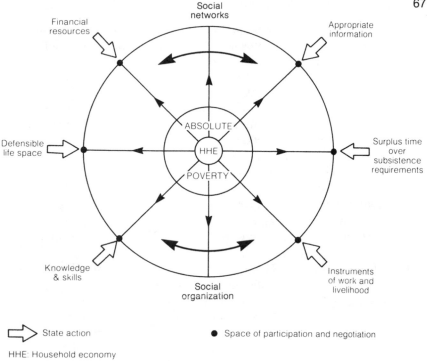

Social
networks

Financial
resources

Appropriate
information

Defensible
life space

ABSOLUTE

HHE

POVERTY

Surplus time
over
subsistence
requirements

Knowledge
& skills

Social
organization

Instruments
of work and
livelihood

⇨ State action ● Space of participation and negotiation

HHE: Household economy

Figure 4.1 Poverty as lack of access to bases of social power

Social power is the power associated with civil society; it is limited by contrasting forms of state, economic, and political power (see figure 2.2). Each form of power is based on certain resources that can be accessed by a collective actor. The state has the law on its side and a monopoly over the legitimate use of violence. Corporations have substantial access to financial resources, the power to shift capital from one place to another, and the power to hire and fire. The political community – parties, social movements, political action committees – has the power to vote, to stage street demonstrations and rallies, and to pressure politicians through lobbying. The power of civil society, finally, is gauged by the differential access of households to the bases of social power.

There are eight bases of social power, the principal means available to a household economy in the production of its life and livelihood.

1 *Defensible life space.* The territorial base of the household economy, defensible life space includes the physical space in which household members cook, eat, sleep, and secure their personal possessions. In a wider sense, it extends beyond the space called "home" to the immediate neighborhood where socializing and other life-supporting

activities take place, chiefly in the context of the moral economy of nonmarket relations. Gaining a secure and permanent foothold in a friendly and supportive urban neighborhood is the most highly prized social power of all, and households are prepared to make almost any kind of sacrifice to obtain it.

2 *Surplus time.* This is the time available to the household economy over and above the time necessary for gaining a subsistence livelihood. It is a function of many things, such as the time spent on the journey to (wage-paying) work; the ease with which basic consumption items such as food, water, and fuel can be obtained; the frequency of illness in the household and access to medical services; the time required for the performance of essential domestic chores; and the gender division of labor. Without access to surplus time, household options are severely constrained. It is the second most prized base of social power.

3 *Knowledge and skills.* This refers to both the educational levels and the mastery of specific skills by members of the household economy. Poor households correctly perceive that education and technical training for at least some of its members are essential for enhancing its long-term economic prospects. They are therefore willing to heavily invest their time, energy, and money in the development of the household's "human resources" (see figure 3.1).

4 *Appropriate information.* This is reasonably accurate information bearing on a household's struggle for subsistence, including such matters as better methods of household production, improved sanitation practices, proven methods of infant care, standard health practices, available public services, changing political configurations, and opportunities for wage-paying work. Without the continuing access to relevant information, knowledge and skills are virtually useless as a resource for self-development.

5 *Social organization.* This refers to both formal and informal organizations to which household members may belong, including churches, mothers' clubs, sports clubs, neighborhood improvement associations, credit circles, discussion groups, tenant associations, peasant syndicates, and irrigation associations. Organizations are not only the means for a more convivial life; they are also a source of relevant information, mutual support, and collective action. They connect the household with the outer society.

6 *Social networks.* These are essential for self-reliant actions based on reciprocity. They tend to increase with membership in social organizations but are not exclusively determined by such membership. Households with extensive horizontal networks among family, friends, and neighbors have a larger space of maneuver than house-

holds without them. Vertical networks, up through the social hierarchy, give households a chance to access other forms of power but
may lead to dependent patron-client relationships.

7   *Instruments of work and livelihood.* These are the tools of household
production: vigorous and healthy bodies (physical strength) and, for
rural producers, access to water and productive land. They also
include the tools used in the household's informal work (bicycles,
sewing machine) and in the domestic sphere itself (stove, pail, kitchen
implements, toilet facilities, etc.).

8   *Financial resources.* These include the net monetary income of households (see figure 3.1) as well as formal and informal credit arrangements.

These eight bases of social power are distinct yet interdependent.
Because all refer to means for obtaining other means in a spiraling
process of increasing social power, they are interdependent. Yet because
they cannot be collapsed into a single dimension such as money, which
mainstream doctrine regards as the principal means for household "empowerment," they are also independent of each other.

Relative access is a measure of the extent to which households command the basic resources for their self-development. Households can gain
greater access to resources in numerous ways. Whatever the method –
and a reallocation of available time resources among household members
will invariably be involved – all households must have *some* access in
order to survive at all.

Conceptually, access may be measured from the center of figure 4.1, a
virtual zero, to the outer rim of the diagram, which represents a non-
quantifiable (theoretical) maximum. Households will be arrayed at different points along each of the eight spokes of the wheel, from virtual zero
to the hypothetical limit. Obviously, no single yardstick can be used to
measure access to the several bases of social power. Rough comparisons
among households can nevertheless be made. In general, increasing access
along any of the dimensions shown will improve a household's condition
of life and livelihood, and therefore it constitutes a measure of genuine
development.

Relative access also allows us to conceptualize a level of *absolute
poverty* consistent with the model's multidimensional view of poverty.
People living at or below this "line" may be unable to move out of
poverty on their own. Nevertheless, the model allows each household to
make its own decisions on how to use its resources for gaining greater
access to the several bases of social power. Most households initially tend
to seek some firm grounding for their activities (in the countryside a piece
of land; in the city minimally adequate housing). Surplus time is often a

second priority, and both may be dependent on households' social networks and participation in popular organizations. Once these "basic needs" are minimally satisfied, however, households may set very different priorities for themselves, pursuing very different ends. At this point, collective action based on shared purposes is likely to yield to more individual-familistic action.

Households' struggles to gain greater access to bases of social power is partly a self-reliant effort and partly a political and therefore collective struggle to put forward *claims* on the state for financial and/or technical assistance. But most households would not want the state to act unilaterally: they want meaningful help, not handouts that would return them to the status of wards. Households want not only to be consulted about but to take an active part in the provisioning of their needs. Figure 4.1 accordingly shows six spaces of participation and negotiation in which households can negotiate solutions to their problems with agents of the state.

Such spaces are not shown for either social organization or social networks, which are power bases of civil society from which the state is excluded. Acting in collaboration with others and beyond the state's reach, households can increase their chances of gaining access to the remaining bases of power. This is indicated by the lateral arrows fanning out from the vertical axis in figure 4.1. The model of (dis)empowerment may thus also be viewed as an empowerment model or, more accurately, a model of *collective self-empowerment*. It is thus not only a model of poverty and deprivation but also a model of how poverty can be overcome and a genuine development promoted.

Even so, there are serious limitations to the prescriptive uses of the model. By making household action central to it, its spatial referent is the microsphere of the locality. It is at this level that households can perceive their interests most clearly and are also motivated and engaged. But the constraints on what can be locally achieved are severe, for poverty is a condition of systematic disempowerment whereby implied *structural conditions* keep the poor poor and confine their access to social power to the level of day-today survival.

In chapter 2 I argued that mainstream economic growth renders much of the population superfluous to the needs of capital accumulation. The constraints on the poor are therefore structural in the sense that the system of power relations that sustains capitalist production also acts to keep the poor disempowered. It fails to provide for the full employment in the formal economy; it fosters a pattern of land ownership that reduces small peasants to a condition of virtual landlessness; it shuts out the "underclass" from effective political participation. To move beyond survival, then, means that the dominant relations of power in society will have to change. This calls for something beyond an increase in access to the bases of social power. It calls for the transformation of social into political power and a politics capable of turning political claims into legitimate entitlements (see chapter 7). Embattled as they are, social democracies of the West have acknowledged household rights to life space and surplus time for a long time. But the institutionalization of these rights was preceded by decades of hard struggle for worker housing, a 40-hour work week, a minimum wage, and child-care services.

The (dis)empowerment model of poverty is both descriptive and prospective. It helps us to look at poverty from the perspective of those who are trying to make ends meet and, if possible, to better their lives. The active center of this effort is the household economy. Poverty in this view is to be disempowered relative to certain bases of social power. But the model can also be turned inside out and so become prospective. Those who are relatively disempowered will want more power and are indeed engaged in a lifelong struggle to improve their situation with respect to one or more bases of social power. The basics of this struggle are shown as the horizontal and vertical dimensions of figure 4.1: life space and surplus time, social organization and networks. Once these bases are minimally secured, households can devote their efforts to the remaining dimensions of social power: knowledge, skills, and information; tools of production; and financial resources.

Of course, as households address these issues they encounter extraordinary obstacles. The very poorest − famine victims, landless rural laborers, women-headed households in the squatter areas of big cities − may simply lack the means to help themselves. They require help from religious organizations, labor unions, and even the state. But even for the less destitute, collective self-empowerment is rarely a spontaneous process of community action: external agents are critically important.

The principal purpose of this model, as was true of the the whole-economy model, is heuristic. Both models must now be used to help us define more precisely the tasks an alternative development must address. I will refer to these tasks as political claims, because this is how they appear in the context of social movements that seek to displace the hegemonic growth model with a meaningful alternative capable of attracting political support.

5 Political Claims I: Inclusive Democracy and Appropriate Economic Growth

Immediately following a 1987 world meeting of nongovernmental organizations (NGOs) convened in London by the World Bank in collaboration with the Overseas Development Institute, a more restricted group of NGOs from only poor countries met to discuss cooperation among nongovernmental development organizations. Its report included a brief statement of alternative-development principles that bear repeating before we launch into a detailed discussion of alternative development. These principles are the closest we can come to a contemporary vision of an alternative development from within the poor countries themselves, as defined by development agents other than the state (Drabek, 1987).

According to the conference report, an alternative development seeks to "reverse such destructive processes as impoverishment" by

- permitting the poor to reacquire "the power and control over their own lives and the natural and human resources that exist in their environment";
- strengthening "their inherent capability to define development goals, draw up strategies for self-reliance and be masters of their own destinies";
- refusing "to compromise on issues related to the social and cultural identity of [poor] societies";
- placing "special emphasis on and attention to utilizing and developing the indigenous efforts, however small, that are promoting self-reliance";
- uncoupling from development processes "all aid which is intrinsically tied to the foreign policies of donor states";
- recognizing "that non-governmental development organizations working with the poor and having an indigenous evolution are important vehicles for change in the development process and [that] support should be primarily provided to them"; and
- recognizing "that all development efforts must have as equal partners

women who have until now borne the burden of the anti-development processes."

These principles, it might be argued, are partly self-serving when they insist on channeling development support "primarily" to indigenous nongovernmental development organizations as "important vehicles of change." And the key role of women and their claim to equality – coming as it does at the end of the declaration – appears to be an afterthought, perhaps a result of intense lobbying by women members of the forum. At the same time, the principles stress the territorial character of an alternative development, greater autonomy over the life spaces of the poor in the management of resources, collective self-empowerment, the importance of respecting cultural identities, and the democratic participation of the poor in all phases of development practice. In these things, the principles reinforce what has been said in these pages.

In this and the following chapter, I present the major components of an alternative development as a set of *political claims* for a shift in the allocation of common territorial resources in support of four broad normative orientations that I believe define the layered meanings of an alternative development. These orientations are consistent, if not identical, with the principles set forth by the 1987 London conference.

Normative orientations suggest a broader scope than the more conventional "goals" of planning discourse. Rather than specific end states to be reached, they are meant to serve as "guiding images" of the desirable directions of social change. In the present case, their overarching intent is to reintegrate the invisible poor with the larger community, and to assert their full rights as citizens in that community. But to put the matter thus does not mean that the poor ought simply to be "upped" a notch or two in their customary subaltern roles. Reintegrating as many as half the population with an existing political community in which, at present, they exercise few rights cannot be done in any meaningful sense unless the systems in dominance – authoritarianism, peripheral capitalism, and patriarchy – are themselves fundamentally changed. This battle for systemic change, which may last for several generations, forms the subject matter of chapter 7. Here and in the next chapter, I look at the principal normative orientations (and their underlying arguments) of political, economic, social, and futures integration that are shaping this battle.

Political integration gives rise to the claims of an *inclusive democracy*, which points to a change in the system of political relations and corresponding institutions. Economic integration gives rise to the claims of an *appropriate economic growth*, which articulates territorial (political) with nonterritorial (market) relations. Social integration gives rise to the claims of *gender equality*, or women's claims for equal rights and social participation, and thus for an end to the prevailing patriarchal order. And futures integration gives rise to claims of *intergenerational equity*, which are the claims of future generations to inherit the physical environment in a condition as good as or better than that which their parents and grandparents received it. Implicit in this claiming is a dramatic change in our dealings with the environment, from short-term functional analysis to an ethic that incorporates the interests of future generations in the management of the physical world.

Political Integration: Inclusive Democracy

Contrary to a popularly held view that democracy is defined primarily by a set of individually held rights, such as the vote or free speech, it is here understood to rest on the legitimate powers of an actualized citizenship or of responsible membership in a politically constituted community. As I pointed out in chapter 1, a political community may be viewed as civil society in its political aspect. Like civil society, it is therefore divided by material and ideological interests. In the present context, the term *community* simply conveys the basic equality of rights and obligations of its citizen members.

Political communities come into existence as a result of a voluntary decision by the members of a territory-based social group to peacefully resolve problems that rise to public attention. This, in essence, is the meaning of the contract theory of political community. Accordingly, all polities are constituted territorially: their authority to address public issues extends only to the limits of their common life space.[1]

Political communities exist at different levels of social formation, and all of us participate simultaneously in several such communities, which may range from the polity of the household through intermediate polities of neighborhood, village, district, city, and region, to the nation and even beyond it, to an incipient multinational and ultimately global community.

Sharp conflicts do, of course, arise within and among these several communities, but conflicting interests are supposed to be played out according to established rules, procedures, and institutions. As a means for resolving social conflicts, the use of physical violence is seen as the breakdown of politics.[2] A political community is a system of political order – the "rules of the game."

---

[1]   For a masterly treatment of the idea of democracy, see Dahl, 1990.

[2]   The state, of course, reserves to itself a monopoly of the means of violence, but it is constrained in their use – at least in a democracy – by human rights and judicial procedure. The willful use of violence by the state, when it happens, is regarded as illegitimate and its agents culpable.

Democracy, in this vision, implies that the people organized into political communities are the ultimate source of sovereign power over their life spaces, which, in the modern era, constitute precisely bounded territories. As the "executive organ" of the political community, the state is sovereign over its territory only by extension of the people's sovereignty. For all of its actions, the state is accountable to the community.

Political community refers to a *virtual* power that becomes actual only when it struggles over and articulates particular claims. We have witnessed such articulations in the case of struggles for national independence (India, Eritrea, Slovenia, Lithuania), for greater regional autonomy (Biafra, Punjab, Southern Sudan, Quebec), for political participation (South Africa), and for local self-government. Women's claims for equal rights within the household economy have transformed the latter into a political arena as well. The state (or the institutionalized power of the patriarchal male) may attempt to suppress the actualizing power of the political community and may succeed in this for a long time, especially where a large proportion of the virtual community is excluded from responsible membership, as is generally the case of the poor and of women. But because state power remains illegitimate without the sovereign exercise of political power by "the people" and must therefore be maintained by force, the virtual power of political community, as recent events in Eastern Europe and the Soviet Union have shown, cannot be indefinitely suppressed.

Responsible membership in a political community requires the transformation of social into political power. The concept of citizenship is thus closely tied to the (dis)empowerment model of poverty.[3] Disempowered people, especially those living below the poverty line – though not only they, because exclusion can rest on other grounds as well, such as racism and sexism – lack the requisite social power to meaningfully exercise their political rights. Because they have no excess social power to convert, their political participation depends, in the final instance, on a solution to their subsistence problem. It is poverty that effectively excludes them from the full exercise of their political rights.[4] At the same

---

[3] It should be clear by now that the concept of who is to be counted as citizen in a democracy is an inclusive one. Following Robert Dahl (1990, p. 129), I shall say that the politically active community, which Dahl calls the *demos*, "must include all adult members of the association except transients and persons proved to be mentally defective."

[4] When the masses of poor burst out of their ghettos and engage in street violence, which the Brazilians call *quebra quebras* (smash ups), it may be a political act, but it is not politics. And although it can bring results, usually of a temporary sort, it should be seen as a spontaneous and highly visible expression of extreme frustration with a political process that turns a deaf ear to popular demands. As Hannah Arendt repeatedly reminds us, politics requires speech. Although some scholars have argued that poor people can only gain attention through violent action (Piven and Cloward, 1979), without spokespersons who can articulate the popular rage, very little if anything will be accomplished.

time, political practice as a form of collective self-empowerment is required for carrying out the continuing struggle against deeply ingrained poverty. As a result of this double relation, inclusive democracy must be viewed as the major claim of an alternative development. It constitutes its political framework.

That inclusive democracy is not the existing framework in any of the world's poor countries can be easily verified. Capital accumulation and growth efficiency appear to work best in a depoliticized environment. As some have argued, they are processes best managed without more than the symbolic participation of organized political communities (Crozier et al., 1975). In many poor countries, participation, if encouraged at all, is treated as primarily a ritual act, while movements that seek to convert their social power into political power are repressed. Over much of the world we find state regimes that vary by degrees from authoritarian to totalitarian and are backed by military force.

Authoritarian states occasionally attempt to gain legitimacy with so-called populist measures, such as urban subsidies for basic foods, public transportation, and low-cost housing. Such measures are not, in themselves, empowering, however. All genuine and lasting empowerment must involve a process of collective action by the poor themselves.

What follows is a discussion of the political claims of an inclusive democracy as part of an alternative development. I am conscious of the Western bias inherent in political theories on which these claims rest.[5] Non-Western political traditions do exist, and for the most part they are authoritarian, as in Iran or Japan, or totalitarian, as in China. Latin America, with its organic-statist models of society, is no exception (Stepan, 1985). Nevertheless, the move toward a democratic form of government is now worldwide.

The political claims for an inclusive democracy focus on three interrelated struggles: (a) strengthening the meaning and reality of political community; (b) devolving effective state power to regional and local levels of governance; and (c) increasing the political community's autonomy over its life space.

The struggle for inclusiveness

This claim relates to the structural conditions that may inhibit the formation of political community, leaving it in a condition of virtual rather than actualized power.[6] Inclusiveness refers in the first instance to formal

---

[5]   For relevant theoretical works on democratic theory, see the extensive bibliography in Dahl, 1990.

[6]   All territory-based communities wish for some degree of autonomy over their life space. In some cases, such as nation-states, this desire for political power has become institutionalized. This is called "actualizing political power." In other cases, involving regional or local power, the struggle continues. Where it is not yet institutionalized, we can speak of virtual power.

or customary criteria of citizen rights. The claim is that, in its simplest form, citizenship is a set of rights and obligations that devolves upon all virtual members of a political community, regardless of gender, age, ethnicity, level of education, religion, or any other criterion that may be used to restrict responsible membership and exclude citizens from the unhindered exercise of their political rights and civic duties.[7]

As important as the formal inclusion in an actualized political community of all who permanently reside within a given territory is, an even more radical claim is for the eradication of absolute poverty. Destitution may well serve as a more effective device for political exclusion than any arbitrary criterion. Poverty, we have argued, is a form of disempowerment, and absolute poverty tends to absorb available household energies in the myriad activities that ensure daily survival.[8]

But the claim for the eradication of absolute poverty, which in this instance rests on the "natural" right to participate in the political practices of civil society, is not to be understood as merely a precondition for inclusion. In an alternative development, the only acceptable way to bring households *above* the threshold of poverty is through collective self-empowerment. Political practice, albeit confined to localities, is, as Aristotle noted, means as well as end; it is one of the ways by which we affirm our common humanity. At the same time, in referring to collective self-empowerment I do not imply that the poor must rely solely on their own efforts. Among the poor, self-empowerment is unlikely to occur of its own accord, just as pupils require a teacher. The teacher teaches, and the student learns. And yet learning can be called a form of self-empowerment. The teacher can be a helper and catalyst who encourages, provides opportunities, tells stories, asks questions, corrects mistakes, and serves as a resource. The actual learning, however, remains the exclusive responsibility of the student.

The crux of the matter, then, is how poor people are to be helped. Genuine empowerment can never be conferred from outside. In the struggle against poverty and for political inclusion, the role of external agents is to provide support in ways that encourage the disempowered to free themselves of traditional dependency. Outside agents working to help the poor to gain a foothold in the city, to reduce unnecessary expenditures of time and energy through the provision of basic services such as drinking water, or to acquire useful knowledge and skills must

---

[7] But see note 3 for Dahl's restriction regarding the inclusion of "transients" and "mentally defective" people in a political community. Children constitute a special case and may be treated as virtual citizens.

[8] Absolute poverty also limits horizons of expectation and has a vicious effect on self-regard. Disempowered people tend to internalize the negative image that the more empowered tend to have of them. For one perspective on this complex issue, see Papanek, 1990.

encourage the poor to overcome their fear of becoming active in the communities in which they live, to acquire a positive self-image, to speak their mind confidently, to identify and support local leaders, and to seek cooperative solutions. The emphasis must always be on learning through collective action. In an excellent statement on the subject, Diego Palma writes, "Community organizations are the place where people learn the praxis of a real democracy, learn to defend one position and to listen to another, to decide together, to divide the work to be done, to set object-ives. It is the place where experiments can be attempted, in small as in large matters, with all the joy and strength of work in solidarity with others" (1988, p. 25).[9]

The responsible exercise of citizenship is learned as people gain cour-age, discover new horizons, and gradually become aware of their rights as members of free political communities (Freire, 1973, 1981). But as long as there is no political space for civic encounter and mobilization – that is, no space for acting as a political subject – citizenship remains merely a virtual power. Relatively few national societies exist in which the basic freedoms that allow for political action – assembly, speech, information, due process, and so on – have become institutionalized to such a degree that they are taken for granted. Most states have, to varying degrees, closed the space of encounter and mobilization by im-posing a fear of informers, police repression, torture, "disappearances," and other horrors.

But fear can be conquered, and the struggle for democratic institutions – for an open political space – will be advanced only when people are no longer afraid of repression (Lechner, 1988). The struggle for an open political space is thus identical with struggles for a political community that can give effective and differentiated expression to its views, claims, and demands.[10]

The struggles for inclusiveness and for an open political space both have in view a restructuring of existing relations of power that limit the options of the poor and prevent their full integration with the larger society. Such attempts will thus be met with almost certain opposition from a political class that stands to lose its privileges and will resort to every possible means including physical violence to prevent such a

---

[9] On the related subject of action research, see the excellent work of Fals Borda, 1985, 1986.
[10] The preceding discussion has greatly profited from a number of Latin America authors who observed that, beginning in the mid-1970s, poor people's movements in São Paulo, Lima, and elsewhere were subtly changing from demand-making to rights-claiming. They were among the first to introduce the concept, now part of everyday language, of civil society in its new sense of a household-centred community struggling for greater autonomy and the first, too, to connect the discussion of housing, transportation, and other infrastructure demands with the broader pro-cesses of democratization. See Moisés, 1981, 1986; Tovar, 1986; and Jacobi, 1989.

restructuring.[11] These struggles promise to be very tough and very long.

There is a tendency to regard citizenship and political practice as a right enshrined at the national level. In fact, however, the specific poverty struggles with which an alternative development is concerned must, in the first instance, be waged locally. It is in the localities where they live that poor people can be mobilized for action. Yet many nations are excessively centralized, which makes it difficult, if not impossible, to address local issues involving the poor. And so the devolution of state power constitutes a second political claim for an inclusive democracy.

The struggle for a restructured areal division of powers

Centralized states are often thought to be strong states. Here I assert the opposite. At issue is the optimal areal division of powers. Excessive centralism – a typical situation in poor countries where virtually all local and regional decisions are run through national ministries – leads, pace conventional wisdom, to a weakening of the state.

This conclusion rests on two arguments. First, highly centralized forms of governance generate administrative pathologies: communication over-load, long response times, filtering and distortion of information, a failure to grasp spatial connections in sectoral programming, and so forth. Contrary arguments that central planning leads to better coordination lack substance (Downs, 1967; Lindblom, 1977). Second – and this is a political argument – centralized states tend to be unresponsive to local needs, and to the needs of the disempowered in particular. They tend to close the political space for civil encounter and mobilization and are likely to perceive the political community as a threat. But a state that is both misgoverned and repressive eventually loses its legitimacy to remain in power. And in the measure that it loses legitimacy, it will be tempted to resort to force. But the application of force swallows resources that might be better used if only the political space were opened up. It also shifts power to the police apparatus and to the military, brutalizes social relations, and encourages corruption. And a state in this predicament is a weak state.

Restructuring the areal division of powers by devolving central functions and resources to regional and local governments thus becomes a central claim of an alternative development. It goes hand in hand with a

---

[11] Some scholars regard a participatory democracy with reserve. One argument declares – in the aftermath of the 1960s outbreaks of popular violence – that participatory democracy renders the polity ungovernable. These scholars would prefer to see less direct participation in public affairs. See Crozier et al., 1975.

revival of political community at these levels and the formation of demo-cratic, responsive, and competent local states, for the longer-term aim here is to create a local (regional) political space that will allow issues of social integration and appropriate economic growth to be resolved through political means rather than bureaucratic fiat, through negotia-tion rather than blind imposition.

Critics on the Left sometimes bemoan the devolution of powers to local governments. They perceive it not only as a fragmentation of political space but of working-class solidarity as well (Markusen, 1987; Slater, 1989). Many Marxists have had difficulties in coming to terms with territoriality and, for that matter, with political (as opposed to revolutionary) practice (Soja, 1989). But as suggested above, centralized states are not necessarily strong states, and essential social reforms may be more readily carried out in an environment of decentralized power. The devolution, spatial fragmentation, and creation of an open political space at local as well as national levels constitute a socially progressive change.

As regards the argument that working-class solidarity will be threatened by the fragmentation of political space, we may well recall the nature of the alternative project. Contrary to the Leninist position that looks to the seizure of state power by a victorious working class, the territorial struggles for an inclusive democracy do not reach out for the state but for political power within a democratic system. And empower-ing the poor in ways that will allow them to be politically effective is a way also of strengthening the political system of which the state is a part.

This is not to argue that class struggle is unimportant. But the organ-ized working class in poor countries is not only very small (typically less than 10 percent of the economically active population); it may also be fearful to advocate an alternative path, where economic growth is "appropriate," and where both social integration and sustainability may reduce the rate of growth by the conventional measure.[12]

In political terms, an alternative development is correctly perceived as a series of interrelated, territory-based struggles involving small and landless peasants and an urban "underclass." These struggles must pro-ceed via class alliances rather than through the one-sided struggles of a self-declared vanguard, which is more properly seen, at least in the

---

[12]   Ray Pahl (1989) argues against the centrality of social class in urban and regional analysis. It is patently obvious that there are many social divisions that are every bit as powerful as class, with gender, ethnicity, and religion in the lead. And these other divisions tend toward the local/regional as the stage for their political practice. For an opposing view from Peru, see Galín et al., 1986.

context of a poor country, as a labor aristocracy largely disinterested in, and possibly even opposed to, an alternative development.[13]

The struggle for accountability and political control over the market sphere

It is an empirically verifiable phenomenon that people wish to protect the environment on which they depend for their life and livelihood. "Protection" in this instance refers to the interest of a political community to extend its authority over the life space or territory that sustains it. At issue is the defense of its borders, the quality of life within its home territory, and the quality of its physical environment. It is this simple idea of protecting what is most intimately connected to one's life that underlies the political claim for greater autonomy over a community's life space – that is, for a greater degree of territorial self-determination.

Poor and rich states alike are losing territorial autonomy as their territories become integrated with the international division of labor and the financial system that makes it possible (Dahl, 1990, part 6). International lending institutions and transnational corporations are able to drive hard bargains whenever demand for foreign capital outruns supply. And regions and localities – subnational political spaces – tend to have even less capacity to impose their will on mobile capital, although relative size is only one variable in their capacity for self-determination. Another is popular support for the state, or the legitimate exercise of its power. All things equal, a strong political community will be more effective in controlling its life space and adaptating changing to external conditions than a weak one.

Two broad conditions influence a community's control over its life space: the state's accountability to the community and the community's ability to assert effective control over the market sphere within its territory.

A strong political community not only requires an open political space in which to mobilize, but it must also be able to hold the state accountable for its actions, for states, it can be argued, exist for no other purpose than to protect and further the territorial interests, or common good, of the community. Whether or not they do so, and whether to the full

---

[13] A powerful movement is currently under way in Latin America to strengthen local governments. For the first time in history, in countries such as Colombia local elections are taking place. In this connection, it may be interesting to note that one of the first achievements of the democratic transition in Brazil was the reconstitution of elected regional (state) governments. On the matter of local government reform in Latin America, see Pease García, 1988; Borja et al., 1987; Grillo, 1988; Herzer and Pírez, 1988; Coraggio, 1988; Rojas Julca, 1989.

extent, cannot be left to their own determination: a full accounting to the people who are sovereign is required. The determination of what will further the common good is not, of course, an easy and straightforward matter. Is economic growth more or less important than environmental quality? On questions such as this, people can and do honestly divide, and when different views are expressed an informed and open debate is the only political way to search for an acceptable solution. Accountability constitutes part of the political process so conceived.

To be accountable means to be obliged to render full and truthful reports to a superior level of authority concerning one's activities and actions. And because democratic theory holds that the people organized into political communities are the legitimate sovereign, it is, in the final analysis, to the people that the state and its agents must give full and truthful reports of their actions. It is this relationship, which many states have unfortunately been able to invert, that allows us to speak of territorial self-governance.

Accountability is a requirement at all relevant levels of territorial governance. This holds true even for the democratic household – conceived here as the smallest political community – whose members are, at least in a feminist perspective, accountable to each other. In actual practice, of course, household members fight over precisely this issue. Are husbands accountable to their wives over how they spend the money they earn? It is a struggle over which households frequently break up. In general, local communities tend to be more transparent and to require fewer formal procedures for ensuring accountability than larger ones that are too complex and remote from people's everyday lives to allow for direct observation of their performance. Relative opacity requires more formal institutions – periodic elections, equal access to information, an inquiring press, and an independent judiciary, among others – to ensure their full accountability.

When all is said and done, however, it is within the local sphere that accountability works best, because it is here that the personal can be observed directly alongside the formal, and here that *who* a person is becomes a factor in judging *what* he or she does. More importantly, people can make connections between public actions and their own conditions of life in ways not possible when issues are posed more abstractly. Nowhere is this more true than for the popular sectors, whose members tend to trust what they see and experience more than what they read (assuming that they can read at all, which many cannot).

If requiring the state to be accountable is one way by which political communities assure themselves of relative autonomy, effective control over the market sphere is another and certainly no less important way. As Amitai Etzioni (1988) reminds us, power relations are essential to exchange relations. And if the disempowered are going to be empowered, and if life space is to be protected against private greed, the power of the corporate economy must be reined in. Social purposes must be made to prevail.[14]

The dominance of neoliberal theory in the 1980s was a step in the opposite direction. Its aim was to depoliticize the economy or, in Alfred Stepan's apt phrase, to "marketize the state" (1985). Chile under Augusto Pinochet was to be the great experiment of this doctrine. Yet oddly, the imposition of lassez-faire did not expand freedoms but brought on 17 years of military dictatorship. By 1990 democracy returned and politics returned with it. Far from destroying civil society, the market radicalism of the "Santiago boys" had contributed to its strengthening.

As regards the role of the state in economic growth, this is nowhere more evident than in the market societies of Korea and Taiwan. States typically intervene over a wide range of issues, including the social and spatial distribution of income and wealth; the regulation of business activities to safeguard health, safety, and the quality of the environment; the provision of services that fall outside the market sphere; and the promotion of employment. In Hong Kong and Singapore – two widely cited free enterprise economies – the state invested heavily in public housing. States also intervene in market pricing by regulating money supply, tariffs, interest rates, and foreign exchange. And they set the economic and fiscal parameters for private business. The minimalist state of neoliberal theory is a fiction.

In the determination of these and similar policies, spatial differentiation is of paramount importance. Indeed, the fine grain of spatial policy may be more important than bucketfuls of good intentions. Whether the matter concerns natural resources, pollution, housing, transportation, or small-business development, the precise nature of the programs and their relative priorities should be a matter for local and/or regional determination, involving full discussion by an inclusive political community that is well informed about these matters. Where the political space is open, it is at local/regional levels that a pure growth-efficiency point of view is least likely to prevail.

Effective state intervention in markets requires a dynamic political community and demands *more* politics rather than less. Contrary arguments in support of technocratic rule are unconvincing.[15] In political discussions, when the time comes to speak out, one's level of formal

[14] Etzioni puts the matter succinctly: "Structures able to limit the *political* power of economic competitors are as important to sustaining competition as is preventing a large concentration of economic power" (1988, p. 257).
[15] See Hall, 1982, for an account of "great planning disasters" in which technocractic decision-making was centrally involved.

education is generally of less importance than full, accurate, timely, relevant, and comprehensible information. Concerning their own conditions of life and livelihood, people tend to be their own best judges.

Conclusion

The political claims for an inclusive democracy may seem quite unrealistic to critics who point to the presumed political lethargy of the poor. But that is precisely the issue in an alternative development: how to break through the constraining conditions of poverty so that the poor may recover their rights as responsible members of a political community. And lest there be any doubts on this matter, it is precisely the claim for an inclusive democracy that is being passionately reasserted in Latin America today. Increasingly, the poor are claiming citizen rights such as for housing rather than a merely temporary relief from their immediate problems. Although demand-making continues in the old mold, the new mold is politically conscious.

Inclusive democracy incorporates a fine areal division of powers; it insists on accountability as a central process and secures an open political space for civil encounter and mobilization. Such a democracy, which includes all potential interests and concerns, will assign a significant role to organized civil society, including the very poor, in the making of public decisions at all relevant levels.

Yet for all that, inclusive democracy requires a state with a highly evolved capacity for both innovation and regulation; a state that will help bring about a new kind of polity, engage in long-range resources and investment planning, and create more equitable conditions throughout its territory; a state that is prepared to defend territorial interests and life ways and to pursue the normative orientations of an alternative development. To act locally is not enough. Alternative development requires a translocal politics.

Economic Integration: Appropriate Economic Growth

I shall approach economic integration in line with the reconceptualization of poverty described in chapter 4. There I defined poverty in terms of a household economy's relative lack of access to the principal bases of social power; here I extend that defining lack of access to include the bases of its productive wealth. For households have a triple aspect: as social formations, polities, and economies. Conceptually, social power belongs to the first, productive wealth to the third.

To treat household economies from a production standpoint, as I propose to do, is more than a question of relabeling the household's activities. Reconceptualizing households has obviously a political purpose. Alternative development calls for a significant shift of resources out of present or potential employments to the poor. Opponents have often argued that this would take resources away from investment in "productive" activities (meaning exclusively production in the market sphere) in order to increase the "consumption" levels of the poor who, as a group, are believed to contribute very little to production in this limited sense, even as they display a high propensity to consume (and conversely, a low propensity to save). I suggest a different position. From a productivist perspective, the poor may be seen as lacking not merely access to bases of social power but also access – and this is much the same thing, although in a different context – to the basic means of production. A shift of public resources from other potential employments to the poor would therefore help to make them more productive not only in the market sense but in terms of the production of life and livelihood within the framework of the whole economy.[16]

Resources shifted out of other employments in the market economy to the household sphere can therefore be expected to generate multipliers in household production and welfare. In other words, even modest expenditures on the productive bases of the household economy are likely to lead to a remarkable improvement in the conditions of its life and livelihood.

In discussing this claim for the economic integration of the disempowered sectors, I consider four distinct aspects: household access to the bases of productive wealth, rural development, informally organized work, and qualitative growth.

Household access to the bases of productive wealth

Whereas gains in social power become available for conversion to political power, an increase in the productive wealth of households converts directly to an improvement in their conditions of life and livelihood. The bases of social power are also a household's bases of productive wealth, or its capital stock. Improving a household's access to these bases therefore increases its capital stock of information, tools, infrastructure, organization, and so forth and, by implication, its productivity as well, so that, for the same expenditure of physical energy, the household will now

[16] Even when accepting this argument, neoclassical economist might object that the productivity of an investment in the household economy of the poor would tend to be lower than in other forms of employment. But this argument is hard to sustain. To measure productivity in social terms, a way must be devised that would allow for its weighing by collective preferences. And this would be a difficult and contentious exercise. In any event, our present accounting systems are hardly adequate to allow for even a rough approximation of an elusive concept such as "rates of return on capital" (Block, 1990).

register greater overall satisfactions in the conditions of its existence (Evers, 1989).

The specific ways by which the household's capital stock can be improved to generate the material basis for an alternative development can be illustrated by two of the cardinal dimensions of our (dis)empowerment model: the quality of life space and surplus time.

Daily life space   Central to all household economies are certain activities, such as sleeping, cooking, eating, cleaning; nurturing small children, the sick, the elderly, and loved ones; securing personal effects and other property; and conviviality. In addition, many poor households raise small livestock and, especially in the countryside, grow grains and vegetables, both for home consumption and for sale. They also carry out from the household sphere a wide spectrum of modest commercial activities, such as selling convenience goods, hairdressing, weaving, tailoring, repair services, food preparation, and so forth.

That all of these activities require some kind of physical space is scarcely an original observation, and from this it follows that where space is inadequate or functionally inappropriate to its purpose the activities concerned will be impaired, conducted less efficiently or effectively, and may even give rise to serious dysfunctions. For example, the space may be both unsanitary and unsafe (Burns, 1970).

Many poor households lack adequate space for even such biological necessities as sleeping. Household members may sleep several to a bed, or take turns sleeping. Limited shelter may be shared with other families who, without housing of their own, are even worse off, giving rise to extreme crowding. Shelter may also give inadequate protection against cold, rain, or dust. Its physical integrity may be threatened by the state for not conforming with legislation. And lacking basic services, it may be insalubrious, resulting in high rates of infant mortality and illness (Jacobi, 1989; Riofrio and Driant, 1989). In short, housing services that are essential for the household economy to function well are often inadequate or absent. Poor households that lack a backyard, sewing room, tool shed, storage area, or other space for commercial use are particularly vulnerable. Many households cannot satisfy even minimal needs if they are exclusively dependent on work performed outside the home. This is especially true for women-headed households, as women are more strongly tied to the home than men (Moser and Peake, 1987; Hardoy and Satterthwaite, 1989).

Surplus time   Where available, surplus time – that quantity of time not needed for basic subsistence that is available to households – can be reallocated from low- to high-productivity activities in either the market or domestic sphere. Surplus time is also necessary, for participating in social organizations and to extend and maintain social networks. One reason that the very poor are characterized by low participation rates and small networks is that they typically lack the time to be active in them. Time-budget studies may suggest whether time is used efficiently or not, but there is no question that poor households, and very poor households even more, are forced to spend a great deal of time on activities that nonpoor households scarcely experience as using up any time at all, such as obtaining potable water or gathering wood and making an open fire in order to cook. A great deal of time and energy is thus "wasted" because water and fuel are not readily available to poor and very poor households. Traveling a distance to obtain basic necessities is a major absorbent time that would otherwise be available for more productive activities.[17] A great deal of time is also spent nurturing babies and small children. Although older siblings can sometimes assume responsibilities in this department, the absence of child-care facilities renders it virtually impossible for the mother to take up work in the market economy. At the same time, men (and women) may spend hours commuting on buses (when they can afford them: more than a quarter of a household income can be spent on transportation to and from work), and if one adds to this the long hours at work (well above the normal eight-hour days of many industrialized countries) or looking for work, no time is left for escaping from the poverty trap.

All these examples make it clear that resource expenditures to improve the housing conditions of the poor and to save time in the performance of household work, and in the provisioning of households, could enormously contribute to their well-being and development. Among other things, it would free the household economy's resources for more productive use.

Rural development

In the poorer countries of the world, rural people still comprise more than half the population, and in the poorest, their proportion may rise to 80 percent and more (Grindle, 1988, table 1). Despite rapid urbanization at more than double the rate of natural increase, the number of rural people in the world continues to grow over a land base that is expected

---

[17] The microeconomics of time allocation on questions related to basic subsistence are illustrated in great detail for Port-au-Prince, Haiti, by Fass, 1988.

to remain nearly constant (World Resources Institute, 1986, table 4.2).[18] This is the first set of facts to remember. The second significant fact is that agricultural production in poor countries has, for the most part, been increasing at rates faster than the total population. And although a rising proportion of this production is intended for export, food production for domestic consumption, except in Sub-Saharan Africa, is also reported to be up (World Bank, 1989a, table 4). Despite this admirable record, the vast majority of poor countries are importing large quantities of food, especially cereals (table 5.1). The third fact to keep in mind is this: despite increases in domestic food production and continuing large cereal imports, an astounding 34 percent of the population in the so-called developing countries are said to be getting insufficient calories for an active working life (World Resources Institute, 1986, table 4.1).[19] For 30 low-income countries with per-capita incomes of less than $400 in 1980, that percentage rose to 51. These figures reflect not only African famine conditions: an estimated half of the population of seven South Asian countries was also undernourished, having a caloric intake insufficient to prevent stunted growth and serious health risks.[20]

[18]   The differences in these rates among Third World countries are dramatic, and the summation attempted here is accurate in a very general sense only. (For details, see World Bank, 1989a, tables 26 and 31.) Quite aside from problems of measurement, which are considerable, international data are not directly comparable, and urbanization rates in particular must be carefully examined. Moreover, not all rural people subsist in whole or in part from agriculture. Aside from other primary activities, such as cattle herding, fishing, the collecting of forest products, and artisan mining, rural people may engage in secondary or tertiary forms of production, and a considerable number either live or work in small towns that may or may not have urban status but that, at any rate, exist in close symbiosis with the rural economy. The rural population in most poor countries tends to increase at rates that vary from 1 to 2 percent. In the higher-income and also more urbanized countries of Latin America, however, the rural population has begun to decline.
[19]   Although the data are for 1980, there is no evidence that the world food situation has markedly improved during the past decade. Given the decline in real incomes, especially for urban workers, the food situation in many countries may actually have deteriorated.
[20]   Such conclusions have begun to be challenged by scholars who contend that the human body makes adaptations to low-energy intake and that the effects of so-called undernourishment are not nearly as serious as portrayed in the literature. In a recent article, Edmundson and Sukhatme write,

> Populations with low intakes have been labelled "undernourished" and those with high intakes have been called "overnourished." These broad classifications have little economic or developmental significance. The statement that many people in the Third World are underfed and therefore physically underactive has no more significance than its illogical corollary that many Westerners are overfed and therefore hyperactive. Both physical and social adaptation occurs in peoples with low energy intakes. Poorer individuals and populations with low food intakes may be small and lean, yet they work long and hard and are extremely efficient at converting food energy into physical work. Those who eat less sometimes work harder.
> The major problem with low energy intakes ... is rather that forced adaptation results

Table 5.1 Summary data on food exports, imports, and production

Income categories	Primary commodities exports (%)[a]	Cereal imports (thousand MT)		Food aid (thousand MT)		Average index of per-capita food production[b]
	1987	1974	1987	1974–5	1986–7	1985–7
Low Income	22	22,767	27,750	6,002	6,677	115
China and India	17	11,295	15,943	1,582	791	119
Other	27	11,472	11,807	4,420	5,886	106
Lower middle income	27	22,000	36,535	1,600	5,338	101
Upper middle income	15	18,589	35,414	328	25	101

a Other than fuels, minerals, and metals.
b 1979–81 = 100
Source: World Bank, 1989a, tables 4 and 16

Taken together, these facts tell us that, to put it bluntly, rural society is becoming a three-tier society of (a) relatively prosperous farmers who are increasingly practicing a modern agriculture with high inputs of fertilizer, pesticides, improved seed varieties, and irrigation; (b) small-scale producers who consume most of what they grow and whose articulation with the market is minimal; and (c) a rapidly expanding group of landless and near-landless workers who, although they still reside in rural areas, must support themselves by working for wages or otherwise engaging in non-farm occupations at whatever wages they can command (see figure 2.1).

As average per-capita cropland is declining toward a projected estimate of less than 0.3 ha by the year 2000 – a decline of about 50 percent over a 25-year span – the number of landless and near-landless workers is growing year by year (World Resources Institute, 1986, table 4.2).[21] Land reforms – that is, those among the many promised actually carried out – would at best only slow down the descent into penury (Hunt, 1984).

Even small peasant farmers, who produce partly for the market and are better off than the rest, can no longer subsist entirely on the land. Members of their households migrate to the city, sending remittances when they can. Many more hire themselves out locally or engage in some form of local trade. Still others, having been forced off their land by expanding agribusinesses and commercial farmers, move out to the "frontier." In many parts of the world, however, the remaining land frontier is closing down, as increasingly marginal lands are brought into production and tropical rain forests devastated. It is difficult for new settlers to gain a firm footing in these areas. Even where rural colonization is actively promoted by the government, as in Brazil and Indonesia, most settlers find that they are unable to subsist on farming alone. After a few years they may end up moving to a nearby city or town, work for their more successful neighbors, or go still deeper into the forest, becoming part-time hunters and collectors.

In most poor countries about one-half the rural population is thus virtually shut out from meaningful participation in development. They have little education; their life expectancy is short; their energy levels are low; and they live precariously on an ever-diminishing land base, with

in marginal reserves which may be insufficient to cope with any additional external stress. (1990, p. 276)

The authors argue that official aid programs should place greater emphasis on malnutrition than on undernourishment, and that the matter is mainly one of nutritional education. These conclusions, however, are very contentious and need to be carefully reviewed before they can be accepted as the new conventional wisdom.

[21] Landlessness is usually documented only in village studies. See the Bangladesh Rural Advancement Committee, 1983, table 5.1.

few opportunities for income generation. They number one-fifth of the world population, or one billion people.

An alternative development must clearly address the needs of the rural poor. "For those who are poor, physically weak, isolated, vulnerable, and powerless," writes Robert Chambers of the University of Sussex's Institute of Development Studies, "to lose less and gain more requires that processes which deprive them and which maintain their deprivation be slowed, halted, and turned back" (1983, p. 168). Although for the most part invisible, the rural poor have a rightful claim to citizenship in their respective countries, and therefore also a claim on common resources. Disempowered they may be, but attention must be paid to them.

But supposing we did pay attention – what then? The problem, we can agree, is a massive one. If one-half the rural population is isolated, vulnerable, and disempowered within a spectrum of social classes dominated by an elite that owns most of the land or commands most of the productive resources, what practical claims should be made? It is one thing to insist on citizen rights but quite another to actualize these claims. In a politics of claiming, ends and means must be considered jointly.

Some initiatives that have been tried and have had at best limited success can be quickly dismissed. Two of these involve relocating the poor: getting them off the land and into large and presumably dynamic centers of urban-industrial growth (the so-called growth-pole strategy) and resettling them on new land. A third failed strategy is integrated rural development (IRD).

The growth-pole strategy   This was popular, especially in Latin America, during the 1960s (Friedmann and Weaver, 1979) and was intended to channel massive investments into the economic infrastructure of selected cities. Fiscal and other incentives (subsidies) would induce "propulsive" industries to locate in the chosen cities and set in motion a dynamic process of economic growth. "Surplus" rural population would be attracted to these "poles" of employment generation and become productive. With less labor available and greater orientation to regional urban markets, agriculture would be modernized (Rodwin, 1969).

The growth-pole strategy was tried, among other countries, in Venezuela, Chile, Brazil, and Korea. But, with the possible exception of Korea, it did not perform to expectations. The official designation of growth poles often became a political gesture. Some of the new urban-industrial centers flourished, but only as long as subsidies continued. Employment multipliers did not occur in the expected volume; what boomed instead was informally organized work, and mostly on a subsistence basis. Although the poor did migrate to the new centers, more often than not they merely succeeded in exchanging one kind of poverty for another. The expected rationalization of agriculture frequently failed to occur, and when it did occur, it tended to be at the expense of the rural poor.

The urban "pole" did offer better access to social services than more remote rural areas, and it afforded greater opportunities for social mobilization. But that very capacity to mobilize in a politics of claiming was perceived as a threat to the status quo. In a number of Latin American countries, beginning with Brazil in 1964, fear of social revolution contributed to the demise of democracy. Urban growth on the periphery, which is what growth-pole policies were meant to accelerate, thus failed also in the important political dimension.[22]

Lacking income-earning opportunities in peripheralized regions, rural migrants continued to drift toward the traditional centers of political power and capital accumulation, creating the megacities of the second half of the twentieth century: Mexico City, São Paulo, Calcutta, Seoul, and dozens more. Nor did growth-pole investment induce development in the countryside. Very few viable regional centers resulted from the strategy. And so the hopes of siphoning off rural poverty and converting it productive labor at selected dynamic urban centers went largely unfulfilled.

Resettlement strategies   These have been adopted in a number of countries. The intention in Malaysia's Pahang Tenggara scheme was to relocate poor peasants from remote and densely populated regions in northeastern Malaysia to a more central location (Higgins, 1976). Similar efforts have been promoted in Nepal's Terai and in the Amazonian portions of Peru, Ecuador, and Brazil. One of the best-known programs is Indonesia's transmigration strategy, involving the relocation of hundreds of thousands of subsistence farmers and their families from overcrowded Java to outlying islands of the archipelago (Douglass, 1985). The idea underlying each of these cases has been to open up new land by clearing tropical forests and providing homesteading families with basic infrastructure and other facilities. But the cost of such programs can run very high. Estimates for Indonesia range from $20,000 to $50,000 per transmigrant household, only a small portion of which actually reaches the settler family. Despite the heavy outlays, most of the resettled population is scarcely better off than before (Sasaki, 1989). Tropical forest soils quickly leach out; indigenous populations in resettlement areas are hostile toward the newcomers; basic services are often not provided or maintained; and markets are far away. Resettlement schemes – with

---

[22] Based on his reading of Chilean politics at the time, Friedmann (1968) proposed a policy of accelerated urbanization. See also the related article in Friedmann, 1988, chapter 6.

Pahang Tenggara apparently the exception – tend to reproduce poverty more than they empower people. They merely shift poverty's location.

A different approach to resettlement is Tanzania's *ujamaa* movement. Over a period of only 80 years, 13 million peasants were uprooted from their traditional lands and resettled in 7,000 more or less improvised villages (McCall and Skutsch, 1983). The intention was to "capture" the peasantry for a state-guided modernization process along "African social-ist" lines. It was as much to bring them under the tutelage of the state as to facilitate providing them services – services that were easier to deliver to populations concentrated in villages than in dispersed settlements (Hyden, 1980). Initially a voluntary program, the formation of *ujamaa* villages was eventually made mandatory, degenerating into a rage of bureaucratic mismanagement. Its results for Tanzania's peasant house-holds have been mixed, even after discounting the pain and hardships imposed on families by forcible relocation. Programs based on the Tanza-nian model were also tried in postrevolutionary Mozambique but only succeeded in alienating the affected population. Even before the civil war that eventually erupted, the government proved unable to extend meaningful assistance to the newly established villages.

Integrated rural development (IRD)   IRD has been touted as a "progres-sive" strategy by its advocates and sharply criticized by others, especially economists (Ruttan, 1975). It is usually understood to be a multisectoral, multifunctional development initiative placed in one of several different locations. Integration is basically a response to the judgment that the rural farmer's poverty stems from a host of problems requiring a package of coordinated responses – from health services to agricultural extension to credit and technology dissemination. As a result of their relative size and complexity, most IRD projects are donor assisted and have been conceived by donors (Honadle and VanSant, 1985).[23]

---

[23]   The context in which IRD projects are implemented has great significance for management and organizational choices. Honadle and VanSant list six contextualizing factors commonly encountered:

1.   IRD projects are often located near international borders.
2.   IRD projects that emphasize food production may occupy a particularly important position in a national policy that gives priority to providing cheap food for urban populations.
3.   IRD projects frequently are situated in an area with a history of political disaffection toward the national government.
4.   IRD projects often impose changes in the local authority structure by introducing temporary arrangements for project management and by using technical criteria to replace traditional patterns of decision making.
5.   IRD projects are often part of a process of decentralization by the national govern-ment.
6.   IRD projects are usually administratively complex. (1985, p. 6)

IRD projects require a complex bureaucracy to administer them as special development enclaves. Emphasis is on the coordination of investments and the delivery of rural services *outside* the regular line agencies of the government. IRD projects are usually funded outside the regular budget and are always under tremendous pressure to show tangible results. The best of them become showcases for the regime in power, but many projects fail to produce the expected results. In quantitative terms, they constitute little more than a gesture.[24]

If the three kinds of strategies discussed have failed to make a significant dent in the problem of rural poverty, what might be done to empower the rural poor with greater chances of success? The starting point, surely, is with what already exists: so-called indigenous agricultural practices (Richards, 1985). Poor peasants are poor partly because they lack access to good land and water. Their agricultural practices represent a reasonable adaptation to this condition. Far from being ignorant, peasants know their terrain fully well. Experience has taught them how to survive under adverse conditions. An alternative development must be built from existing peasant knowledge and skills.

In a review of rural development projects, Robert Chambers (1987) draws five major lessons for the achievement of a "sustainable livelihood": (a) to follow a learning approach in development, (b) to put people's priorities first, (c) to secure people their rights in and gains from development, (d) to reach sustainability through self-reliance, and (e) to ensure the high caliber, commitment, and continuity of the technical advisory staff. Overall, he argues for working *with* people. He conceives of development as a cooperative endeavor in which poor people's priorities come first, for it is clear to him that projects must continue when the experts leave. New practices have to be acquired by the poor themselves. It is precisely this learning that is part of the empowerment approach.[25]

The projects reviewed by Chambers deal with nutrition, watershed management, agroforestry, and the like, yet we have seen the necessity of providing off-farm employment for the landless as well as for many others whose labor time is not fully absorbed in farming and household activities. The peasant farm family exclusively dependent on farming is

---

[24]   IRD projects are in some ways the rural equivalent of New Town strategies of metropolitan deconcentration. Architects love New Towns because they allow them scope for innovative design. Rural developers are enamored of IRD projects for much the same reason. But the cost and complexity of New Towns, as of IRD projects, are such that they can rarely achieve their ostensible objectives, which reach beyond the individual project to larger policy issues – metropolitan deconcentration in the one, the betterment of rural life in the other.

[25]   On the social learning approach to rural development, see Korten, 1980; for a more general discussion of the theoretical foundations of social learning, see Friedmann, 1987, chapter 5.

becoming a rarity even in the heart of rural Asia and Africa (Bangladesh Rural Advancement Committee, 1983; Ngau, 1989).

Merilee S. Grindle puts the matter well when she writes,

> Agricultural pursuits are only one among many income-generating activities that sustain rural households and communities. Most studies of rural areas and their development have taken as axiomatic that rural households have access to land and that they generate the major portion of their income from the land. Increasingly, this view distorts what occurs in vast numbers of rural communities throughout the Third World. Landlessness and declining employment opportunities in agriculture are increasingly marked.... Off-farm employment cannot be considered a residual source of income for rural households; for many, it is their principal means for meeting their economic needs. (1988, p. 171)

Chambers's agrarian approach must thus be placed in a context of town-centered development. Its broad aim would be to improve the physical infrastructure of rural towns, link towns and cities through improved transportation services, and make credit available to rural businesses that want to establish themselves in the town center. It would also mean using the town as a center for rural cooperatives, health and educational services, and rural extension, and would further suggest the possibility of an integrated political-administrative system at the district level in which rural and urban interests are conjoined.[26]

Finally, adopting the right macropolicies at the national level such as pricing, policies and subsidies, is of crucial importance. As the authors of a 1988 survey of successful rural development projects have observed, "Small changes in government policies can have profound influences on development." In Ghana, for example,

> maize production tripled between 1983 and 1984 after the price of the crop was tripled and the currency devalued. In 1984, Zambia increased the real price of rice by 12 percent, and the amount marketed rose by 55 percent. In 1970, Malawi lost $7.8 million to cereal imports, yet by 1983 it was a net exporter, earning $5.5 million. Prices paid to Malawian farmers for maize have been some 30 percent above world market levels and the nation has one of the most open markets and least overvalued currencies in Africa....
>
> Sustainable development cannot be achieved, except in isolated cases, without policy reforms.... Tax policy, land reform, land-titling, credit policies, commodity prices, tariffs, utility rates, social security programs, low-income housing policy, consumer food subsidies, and labor legislation

---

[26] A town-centered approach such as this – called "agropolitan development" – was suggested as early as 1975. See Friedmann, 1988, chapters 8 and 9.

each have at least as much effect on environmental conditions as environmental policy does. (Reid et al., 1988, p. 29)

This admonition serves as a reminder to which I will return in chapter
7 – that local development successes are not independent of the structural
conditions that constrain them. A politics geared to structural change at
national and even international levels must be part of an alternative
design.

Informally organized work

Informal work, as it is called, is a highly vulnerable activity; nevertheless,
it is of crucial importance for an alternative development. In countries
where neither unemployment insurance nor social security exist, and
where the formal economy is quite unable to productively absorb the
growing increments of urban labor resulting from natural increase and
migration, informal work is one of the principal sources of income for
the poor. From one-third to two-thirds of the economically active urban
population in poor countries may be engaged in as "informal work."[27]
Informal activities occupy a shadowy niche in the economy of metropolitan centers, intermediate cities, and small country towns. They are
both visible and not: visible in the street economy but invisible in a
variety of other ways – the most important of which is because they
don't show up in official statistics. Like so much of poverty and oppression, they are tucked away in the kitchens of bourgeois households,
alleyways, working-class suburbs of large cities, and crumbling innercity
tenements. It is not surprising therefore that, despite voluminous
literature on the subject, precise definitions of "informal" don't exist
and that, in fact, a good deal of scholarly controversy exists about the
matter.[28] In this study I avoid the language of those who refer to

---

[27] Aggregate statistics on informal work are generally not very meaningful. To begin with, there
are certain conceptual problems with "informality". Also there is no clear demarcation between
unemployed workers and those engaged for varying periods of time in informal work. There is the
problem of long hours of work for very low incomes (low labor productivity, self-exploitation), of
the nonreporting of income (the initial basis for the classification), and of the frequent oscillation
of casual workers, especially women, between being unemployed – not earning any money – and
working. For instance, a female vendor who falls ill, gives birth, cares for a sick child at home, or
loses her investment because of theft or her inability to collect on a debt, may drop out of the
informal economy for a period of time. Is she to be counted as employed? Are professional
beggars to be counted? And what of drug sellers and gamblers? Still, despite measurement
problems, the informality tag continues to be widely used to circumscribe a world of life and
work that Diego Palma (1987) calls *el mundo cotidiano de lo popular*–roughly, "the world of
popular everyday life."
[28] For one of the best critiques, see Breman, 1985.

informal activities as a "sector," as if the world of popular everyday life were an internally integrated compartment of production. Instead, I describe informality as work performed in the context of highly fragmented (and segmented) urban labor markets.[29] Informal activities articulate the subsistence economy of households with the accumulation regime of capitalist production, where they constitute one of the principal sources of "surplus value" or profits. It is this intertwining of subsistence and accumulation that contributes to the fuzziness of the concept. Among other things, the volume of informality tends to vary with fluctuations in the business cycle: the greater the economic crisis, the larger the amount of informal work, albeit at very low rates of return. In the world of popular everyday life, informality serves households as a cushion against hard times.

It may be useful at this point to introduce a typology of informally organized work consistent with our attempt to delineate a major dimension of the world of popular everyday life. This will allow us to specify the leading characteristics of informality. (See table 5.2 for relevant details.)

1.   Informal work is mainly for the domestic market. Nonmarket activities, such as building one's own housing and doing domestic work, are excluded from the typology.

2.   Particularly striking is the enormous heterogeneity of informal work. Whether we consider its customary *location* (household, street, employer's household, rented space, or construction site), the mode of its *organization* (unpaid family labor, self-employed, wage labor, microenterprise), its *market relations* (subcontracting, hustling for customers, domestic service, the mediations of a strawboss who recruits a gang of construction workers for a specific job), or its *sectoral affiliation* (artesanal work, manufacturing, personal services, domestic services, retail trades, construction, transportation) informally organized work is clearly not a single "sector" or even a single labor market. Whatever theoretical coherence it may have derives from its role in the subsistence economy of households.

3.   Informal work is typically organized on a very small scale. Many informal workers, especially in the street economy, travel on their own. Sometimes they work in pairs. Unpaid family labor is used extensively, both in the household economy and on the street. This may raise the

---

[29]   In the following discussion of informality, I exclude activities which, like drug peddling, prostitution, smuggling, and most forms of organized gambling contravene against specific legislation and are declared criminal. Although an important source of income and employment for many, the claims of an alternative development do not extend to them.

Table 5.2   Types of informally organized market-oriented activities

Activity	Examples	Customary location	Persons per operation
Home industry	Manufacture: food for vending, handicrafts, clothing Services: washing and ironing Trading: retail	Own household	Predominantly female, including unpaid family labor (1–3)
Street economy	Trading: food stalls, vending Services: shoeshining, portering, transport, entertainment	Street: ambulatory, but also fixed location	Both men and women, (1–3), including some unpaid family labor
Domestic service	Maids, cooks, gardeners, nannies, chauffeurs	Employer household, including live-in arrangements for some staff	Several per high-income household, both men and women
Microenterprise	Manufacture: shoes, tailoring, metalworking Services: electrical and radio repair; plumbing; car repair	Rented space, but may also operate out of own home	Owner-manager, plus several employees (fewer than 10; average 3 to 5)[a]
Construction work	Day laborers, bricklayers, electricians, carpenters	Onsite	Individually recruited for specific projects

[a]   A category of "small industry" with an employment of 10 to 100 workers should be treated outside the framework of informality. See Carr, 1981; Little et al., 1987.

total number of workers per operation to two or three. Microenterprises may use up to ten workers but have an average of fewer than five.[30]

4.   Women workers constitute a variable percentage in informal work. They clearly dominate home industry, have an important role in retail trading, and are strong in domestic service. In most countries they probably account for more than one-third of all informal workers.

5.   Most informal workers work for wages, by the hour or for piece-work. Some sell commodities that may be factory-produced (some of these may be smuggled into other countries) or, like cooked food, prepared by the seller either at home or on the spot. Others, such as

[30]   In his brilliant study of food vending "enterprises" in Port-au-Prince, Simon Fass found that 14 percent of his sample had no employees; 29 and 21 percent employed one or two, respectively; and 15 percent had more than three. Still, as small as these enterprises are, the overall economic impact of food vending in Port-au-Prince is considerable:

shoeshine boys or bicycle repair people, sell specific services for a price that may be fixed or negotiated, depending on client or location. Wages for informal work tend to be below the official minimum level, especially when the employer's business is itself informal. In general, women get 60 percent or less of men's wages. And, of course, most family labor is unpaid.

6.   Some informally organized work is directly linked to large-scale modern manufacturing that is subject to government regulation, unionization, and taxes and is also the principal beneficiary of state subsidies.[31] Formal producers may contract with informal workshops (e.g., in clothing manufacture – the notorious "needle trades") or sub-contract for parts that are subsequently factory-assembled (e.g., shoes, electric motors).

Contracting work to informal businesses creates advantages of flexible production for the firm letting the contract. It lowers production costs and shifts a large part of the uncertainties of the business cycle onto the small and unprotected producer. The bulk of informal work, however, is not so linked into the accumulation economy but remains precapitalist or, at most, protocapitalist in its organization. In either case, the primary objective of the activity is not accumulation but subsistence.

7.   Most informal workers put in many hours for very small returns. In Lima, for instance, one-fifth of all informal wage earners work between 49 and 60 hours, and 23 percent work for more than 61 hours per week (Chávez O'Brien, 1988, table 20).

8.   Different types of work appear to have different educational thresholds. In La Paz, Bolivia, for example, women in home industries and domestic service average little more than three years of schooling, but in "semi-entrepreneurial" activities, which are largely coincident with our microindustry category, women have nearly eight years of education on average (Ardaya, 1986, table 20).[32] Even for ambulatory street

---

There are approximately 23,000 enterprises in Port-au-Prince that prepare and sell cooked meals. These vending establishments employ 74,000 people, the vast majority women, or 11 percent of the urban labor force. One thousand enterprises operate in industrial areas, employ 3,200 individuals, and each day serve meals to 50,000 industrial workers, 75 percent of whom are also women. They spend 30 percent of their daily wages for the meals. In other words, the nutrition, health, and labor productivity of an important share of the city's working women, both those who sell and those who purchase meals, are greatly influenced by the efficiency of vending operations. (Fass, 1989, p. i)

[31]   Of course, to be subject to labor and tax legislation does not mean that a given business will conform in all respects. For instance, it may bribe government inspectors, keep two sets of books, and so forth. These practices, however, do not qualify the business to be counted as an informally organized activity.
[32]   Men had substantially higher education thresholds than women for the first two occupations but showed only 7.3 years of schooling in the third.

vending La Paz women traders – the vast majority being Aymara-speaking – require a rudimentary knowledge of Spanish as well as numeracy, a minimum of working capital, and connections to the wholesale suppliers whose products they carry through the vast marketing networks of the city. These prerequisites are difficult to meet, and other kinds of work, particularly microenterprises tend to have even more stringent conditions. A large proportion of informal workers would therefore have to be called skilled.

9.   Among the most important requirements, in addition to personal networks, is start-up and working capital. Food vendors in Haiti, for example, require a basic capital stock that varies systematically with scale of operation from $97 to $430, figures that compare to an average urban household income estimated at $148 per month, or $62 per worker income. The requirement for working capital is even larger: the biweekly expenditure for commodities ranges from $121 to $695, and for wages from $2.80 to $53.15. Because most customers eat on credit, paying off their debt only when they themselves get paid (usually every two weeks), it is clear that a considerable amount of capital is required to get into the business (Fass, 1989, tables 4, 6, and 8; p. 26). Most of this comes from years of personal savings (over as much as a decade sometimes); smaller amounts may be lent by family, and still smaller amounts by "friends" and commercial money lenders, often at very high rates of interest.

10.   Informally organized work is embedded in a matrix of social relations that are essential to its success. Unpaid family labor contributions, friendship and trust, and patron-client relations are necessary and common attributes of informal work. The high degree of existing market involution means that most street work and microenterprises generate only slim profits. The streets and alleyways of the city are the world of Adam Smith's perfect competition, where sudden price changes in one establishment ripple through a street of small (and similar) food stalls or a line of automobile repair shops. But workers perceive this not as competition but as cooperation. It is to one's nearest neighbor that one turns for carefully gauged doses of help that are expected to be reciprocated in the near future.

Most informal work would cease to exist if it were regulated in ways similar to formal work. Because they vitally depend on informal relations in production for their income, poor people tend to resist attempts to control informality. Were a government nevertheless to succeed in bringing informal work under its scrutiny, as governments repeatedly attempt to do, the most likely result would be an increase in the underground economy. Some advocates of formalization, such as Hernando de Soto (1989), prefer to see informal workers as the vanguard of an entrepreneurial capitalism. Undoubtedly, some microentrepreneurs fit de Soto's description. But the vast majority of informal workers, in Peru and elsewhere, are not vanguard capitalists but unpaid family workers, wage earners, street vendors, day laborers, servants, and repair people. Poor people are sometimes ingenious in the ways they earn money. A street artist in La Paz bites off the heads of lizards to amuse the crowd. But that is not the kind of work that Enrique V. Iglesias, president of the Inter-American Development Bank, has in mind when he writes,

> It is time to stop wringing our hands and start recognizing that in recent years the informal economy has been growing three times faster than the formal one – 7 percent versus 2 percent annually – and that nearly 30 percent of the region's economically active population is employed there. Thanks to studies in Colombia, we now know that $1,000 invested in a productive microenterprise can generate one job, that $400 invested in a service microenterprise can generate another, while $10,000 must be invested in a formal sector manufacturing firm to accomplish the same thing. In a time of scarce resources and high unemployment, we can no longer afford to ignore the informal sector. (1989, p. 42)

Iglesias is an economist and consequently sees informal work as a "sector" of the national economy. He talks about microenterprise as though these one-, two-, and three-person operations were firms like any other in the capitalist economy. But for the most part they are not (see note to table 5.2). To a very large degree, informal work forms part of the subsistence economy of cities. Despite Iglesias's statistics, it is largely invisible, unmeasured work. It is, so to speak, *in* the market but not *of* it. And its political claim is not to be absorbed into the formal economy under the stern, observant gaze of the state, but something else.

From the perspective of an alternative development, what then are the political claims with respect to informal work? In view of the delicate balance between the costs and benefits of informality, the question is a difficult one. Informal workers, such as domestic servants, are indeed "superexploited," but at least they are earning something toward their livelihood (Chaney and García Castro, 1989). Street vendors are continuously hassled by the police and may lose much of their stock in the process. But formalization imposes government controls and taxes and may raise the cost of doing business to prohibitive levels. Formal business is divided over the issue of informality. Manufacturers prefer to deal with an "unorganized sector" that can absorb the uncertainties of doing business and provide cheap labor under hypercompetitive conditions. Finally, the state sees informal business as a means of keeping the social peace but is worried when so many enterprises remain uncounted and constitute, so to speak, an anarchistic margin beyond its reach.

In the context of this politics of informality, five claims can be advanced.

1.   To recognize in public policy the central importance of informal work for the survival economy of poor urban households.

2.   To stress popular education, both to upgrade educational levels generally, especially for women, and to teach specific skills and trades that will be useful in the search for work within informal labor markets. Technical education and training (such as repair of electronic equipment) should be emphasized in apprenticeships.

3.   To promote the organization of informal workers into interest associations, syndicates, cooperatives, and solidarity groups.[33] By organizing, informal workers gain access to bases of productive wealth and social power: the power to negotiate, to bring political pressure to bear on the state, to receive credit, to make bulk purchases at a discount, to rationalize marketing, to prevent dog-eat-dog competition, to establish self-regulating standards of work, and more. Informality and interest organizations appear to be compatible arrangements; 4,000 such associations exist in Lima's Villa El Salvador district alone. And to quote Chilean sociologist Diego Palma once more, "The practice of popular base organizations is embedded in a communal and democratic experience that serves as a school where poor people practice and learn the values that characterize the alternative project" (1987, p. 98).

4.   To make available adequate credit, management advice, marketing assistance, on-the-job training, and other forms of assistance when it is requested by organizations of informal workers. These organizations are the best channels for directing aid not only to the most needy but also to those who are prepared to collaborate with the aid giver in appropriate programs.

5.   To regulate, to the extent possible, serious abuses of informal work arrangements, such as child labor and extreme hazards to health.

This brief agenda seeks to secure a place for informal work in an alternative development. In the course of this development, informality will gradually acquire certain "formal" characteristics. It is not an end in itself but a necessary step in a development sequence whose ultimate aim is the collective self-empowerment of poor households.

Qualitative growth

Economic growth has clearly got to be part of an alternative development. Without sustained increases in the per-capita value of production,

---

[33]   Solidarity groups are small groups (usually fewer than a dozen members) that come together to receive a rotating loan for which members are collectively responsible.

there can be no development in the long run. But the dominant measure of the value of production – the GNP and its component accounts – is deeply flawed, most seriously by failing to take into consideration the social and environmental costs of production. So the statement above must be amended: without increases in the per-capita value of production, *net of social and environmental costs*, it becomes impossible to speak of a genuine development.

Let us assume now that the necessary adjustments in economic accounts have been made – admittedly a daunting task. Would we then have all the instruments needed for an alternative development? The answer is that we would not, and for three reasons: (a) we would want additional information on the social and spatial distribution of income; (b) we would want to disaggregate economic accounts to smaller territorial units of governance – the cities and regions – of a country; and (c) we would want information that would allow us to assess the quality of the actual economic growth.

That income distribution is a basic, albeit insufficient, datum for policy analysis should be obvious, although governments are reluctant to encourage research along these lines, and the subject has received only cursory examination since the mid-1970s. Similarly, it is only common wisdom to disaggregate economic accounts to those territorial levels at which most development decisions are made or, rather, at which they *should* be made so that variable local conditions can be taken into account. And a spatially articulated view of economic growth would also be desirable at the national level, although, again, resistance to national policies for urban and regional development has been considerable, which is an astounding fact if one considers that the governments of poor countries are responsible for more than one-half of all investments and for virtually all the social infrastructure in their countries (Friedmann, 1966). It is to the quality of growth, however, that I want to give closer attention. A less familiar subject, it is presented here as a major political claim of an alternative development.

But what are we to make of quality in this context? And why aren't measures of the market value of production sufficient? The second question is more readily answered than the first. Analysis has shown that correlations of per-capita income with certain social and environmental indicators are considerably less than perfect. If we survey a large number of countries, years-of-education and life-expectancy indicators will tend to increase with rises in per-capita income. But the same is obviously not true for any individual country. And correlations vary a good bit from one indicator to the next. In some instances they may even be negative: urban crime rates, for example, tend to go up with rising income, while indicators of environmental health decline. The same would be true, only more so, for disaggregated variables at urban and regional levels. Per-capita income is thus an unreliable guide to the qualitative impact of growth. If we want the quality of life and livelihood to improve with economic growth, we need to concern ourselves with a wider range of data than merely the (inflation-adjusted) value of production.

To return to our first question, What, in the context of this discussion, is quality? To start with, there can be no such thing as a single, aggregate measure of quality derived from a system of social and environmental accounts. Efforts to construct such an index have been strongly criticized (Ghai et al., 1988). The crux of the debate concerns the weighing of the variables that enter a composite index. With no evident reason why each variable should carry the same weight, any other weighing system would be equally arbitrary, for quality is ultimately a matter of individual and collective judgment, and who is more qualified to pass such judgment than those most directly affected by the decision? The importance attached to quality-of-growth indicator is thus a political matter that must be decided in open and public debate.

A second critical point is of a more practical nature. Measures of quality are intended to improve the making of relevant policies, and for this purpose disaggregated measures are considerably more useful than an artificial construct whose ultimate meaning is far from clear, even to experts.[34]

The quality of economic growth, we conclude, consists of a set of variables that, directly or indirectly, affect household access to the bases of social power and productive wealth. Eight such bases were identified in chapter 4, and for each of these several variables will be needed to measure the quality of growth. Taken together and territorially disaggregated to the relevant political communities, a set of indicators constructed along these lines can tell us a great deal about the local conditions of life and livelihood, especially for disempowered households.

Enormous difficulties have to be overcome to make such a system of social and environmental indicators operational. In 1981 a UNESCO symposium issued 12 recommendations that addressed some of these difficulties. The recommendations include

1   Simplicity of measures must take priority over comprehensiveness....
2   Greater attention must be paid to disaggregated indicators....

---

[34]   In a major review of the use of social indicators in the United States, Judith Innes de Neufville states, "Do not worry about developing integrated, logical systems of social indicators of producing official social reports. The technology and understanding are not yet available to do the former; the latter are not nearly as likely to have an impact as is the publicizing of indicators in conjunction with discussion of policy questions, which for political reasons social reports cannot do" (1984, p. 108).

3   ...[N]o universal system of indicators is practicable.
4   Construction of complex composite indicators is impracticable....
5   The usefulness of any set of indicators depends ultimately on their credibility.
6   Any set of indicators, if it is to be useful for development planning, should meet four criteria: (i) clear definition, reflecting theoretical understanding; (ii) quality – appropriateness to local needs; (iii) currency – timeliness is essential; and (iv) availability – to development planners and analysts.
7   It is important for individual countries to exercise careful selectivity in developing needed indicators with scarce resources. Such indicators should be designed to monitor key programs, so that their relevance to the policy process is ensured. (UNESCO, 1984, pp. 119–20)

It is clear from the context that the UNESCO experts thought of social indicators as primarily an instrument devised by and for planners in the national bureaucracy rather than as information to feed public debate in an open political process.

Given the lack of reliable information concerning the quality of economic growth, especially at local and regional levels, it is essential to mount appropriate research (Mathew and Scott, 1985). But the elaboration of complex systems of social and environmental accounts is unnecessary for the purpose at hand; it may even be counter productive (but see Miles, 1985). The object would be to keep indicators simple, and in the real world, indicators are tools of political advocacy whose very elaboration constitutes a political act.

To quickly recap, political claims for *appropriate* economic growth involve a major shift in policies to directly benefit the excluded sectors of the population. They press for the devolution of decision-making to localities and for the opening up of debate about what ought to be done, and with what means, to improve the conditions of poor households and, more specifically, to improve their access to the bases of productive wealth. An alternative development, as Robert Chambers asserts, must be *with* people rather than *for* them. It must listen to their priorities, involve them in program design, and enlist their active support in implementation. There will be political differences in the community and they will require time to work out, for people's livelihood is at stake, and passions are likely to flare. But an alternative development should not be pressured into producing quick or flashy results. It works close to the ground in a "transactive" style of development, emphasizing mutual learning.[35]

A major emphasis must be put on developing a rural infrastructure that links the relatively isolated areas of peasant farming with domestic

---

[35]   For further discussion of this point, see chapter 7.

markets and assigns a key role in this to district centers. Off-farm employment also needs to be promoted, especially in rural towns and cities, where attention must be given to creating favorable conditions for informal work. The organization of informal workers into associations, cooperatives, and syndicates, is to be encouraged.

Simple indicators of household access to the bases of productive wealth are preferable to composite measures, and the injection of indicators into political debates at different territorial levels is probably the best way to proceed. This conclusion, of course, requires the simultaneous expansion of democratic practices and a form of transactive planning that together constitute the backbone of an alternative development.

6 Political Claims II: Gender Equality and Sustainability

Social Integration: Gender Equality

In chapters 3 and 4, in which households were central to the analysis, we stopped short of analyzing households' internal structure to arrive at theoretical and practical concepts for an alternative development. Households thus remained a "black box," even though we noted that, in addition to being an economy, households must also be looked at as a polity, the smallest territorial unit exhibiting political behavior. Here I analyze the internal workings of households to show that an alternative development must incorporate the structural inequality of gender that is rooted in household relations and is a major source of the tensions and struggles within the household.

Like other territory-based communities, households have a hierarchy mediated by gender, age, and kinship and are based on contractual relations, explicit in the case of marriage and implicit in cohabitation. The state acknowledges households as significant and relatively stable social units and recognizes a de jure "head" of household, usually the man, who thus assumes the right to speak for the household as a whole. Still, many women act as household heads de facto. Their husband may have died, or divorced them, or moved to the city, or simply have left them. In practice, then, women are frequently in charge of feeding and raising their children and otherwise ensuring the household's integrity and fulfillment of its obligations to the larger community. Worldwide, it has been estimated that from 30 to 40 percent of urban households are headed by women. The actual percentage tends to vary a good bit, being relatively high, for example, in Latin America and the Caribbean region and lower in the Arab countries of the Middle East.[1]

[1] It is virtually impossible to get precise figures on women-headed households, in part because of a lack of common definition and the prevalence of male de jure headship even where the man is absent from the household. Bunster and Chaney (1985) put the worldwide figure at one-third. Presumably, the percentage of urban households would be higher. A recent estimate for urban black families puts the proportion of women-headed households in the United States at 44 percent (cited in *National Perspectives Quarterly*, 1990, p. 23).

Following the political hierarchy within households, tasks are divided according to custom by gender as well as by age in a rough allocation of labor time among the principal domains of social practice:

- In the domestic economy of the household: raising subsistence crops, preparing food, cleaning, caring for the sick, organizing celebrations, and so forth.
- In civil society: interhousehold relations, family, community, and religion (temple, mosque, and church).
- In the market economy: "formal" and "informal" work, cash crops, and cooperatives.
- In the state: school attendance, and obligatory military service.
- In the political community: participation in social movements, political parties, and labor organizations.

Within broad parameters, however, the customary division of labor is not inflexible, as households shift around resources, depending on changing circumstances such as births, deaths, illness, aging, migration, household composition, the seasonality of agricultural work, economic crisis, and opportunity. Another variable is the struggle internal to the household itself for a more equitable division of labor. This struggle lies at the core of household politics and renders households far less of a corporate entity than they are frequently believed to be. As Amartya Sen has suggested, households display both cooperative and conflictive behavior, usually at the same time. He proposes a bargaining – that is, a political – model to account for it: "The members of a household face two different kinds of problems simultaneously, one involving *cooperation* (adding to total availabilities) and the other *conflict* (dividing the total availabilities among the members of the household). Social arrangements regarding who does what, who gets to consume what, and who takes what decisions can be seen as responses to this combined problem of cooperation and conflict. The sexual division of labor is one part of such a social arrangement, and it is important to see it in the context of the entire arrangement" (1990, p. 129).

Although the gender division of labor (and the time devoted to various tasks) is one of the principal issues with which intrahousehold political conflicts are concerned, there are others as well, such as sexual relations, control over income, property rights, children's education, and women's participation in civil and political affairs. Although women-headed households may be less afflicted with these struggles, they, too, may have to reckon with the attempts by male members of their extended families, such as eldest son, brother, or uncle, to control their households, their property, and, more specifically, their own lives.

Despite a certain lability of gender roles, the overall structure of household relations throughout the world openly discriminates against women and keeps them in a state of permanent subordination vis-à-vis males, both inside the household and in the wider public domain. The result is women's double disempowerment as members of poor households and, within the household, by virtue of their sex.

The condition of women

In her book *Maternal Thinking* Sara Ruddick argues that "in any culture, maternal commitment is far more voluntary than people like to believe" (1989, p. 22). If by this she means that maternal roles are not biologically determined, that is probably the case. But it is also true that women's biological role as mother reinforces their subordination within the patriarchal order. The high fertility rates still prevalent in most poor countries set strict limits to women's sense of who they are and the survival strategies they can adopt (Benería, 1980). In China, where the state has succeeded in lowering the fertility rate to fewer than two children per reproductive-age woman, and where, at least in cities, extensive provision has been made for subsidized day care, the biological constraint on women's opportunities has begun to be lifted. But China is still very much the exception.

It is their frequent pregnancies, abortions, and births, combined with infants' dependence on their mothers for nourishment (breastfeeding may continue for years), that effectively tie women to the domestic sphere. In most poor countries and particularly among poor people in the countryside, it is not unusual for women who start having babies in their teens to give birth to six, seven, and even more children during their lifetime.[2] Because some of these children may die before reaching adulthood, high fertility rates have often been thought to represent a safeguard to ensure a small margin of survivors.[3] With the declines in mortality rates, however, that margin has been increasing, and in many cases today high fertility represents an instance of social lag. This is not to deny the economic value of surviving children, who generally help out with

---

[2] The average fertility of Kenyan women in 1987 was 7.7 live births. In rural Kenya it was probably higher still. By contrast, in China, extensive campaigns to reduce fertility has yielded a decline to 2.4 (the official goal is a fertility rate of one). But China is an exception among poor countries. Even India's fertility rate, although lower than those of most poor countries, still hovers around 4.3 (World Bank, 1989a, table 27).

[3] According to World Bank statistics (1989a, table 32), infant mortality per 1,000 live births in 1987 was as follows: China, 32; India, 99; low-income countries other than China or India, 109 (with a high of 169 for Mali); lower-middle-income countries, 61 (with a high of 116 for the Yemen Arab Republic); upper-middle-income countries, 50. By contrast, high-income countries (other than those in the Arabia peninsula) went from a low of four in Norway to 15 in Japan.

domestic chores, earn money beyond their keep, and provide an insurance against their parents' infirmity and old age. Whatever the reason for high fertility, it is women who must live through these cycles of birth and dying, raising the survivors with whatever means are at hand.

Rearing large families under the circumstances most poor women encounter, especially women who are heads of household, places extraordinary burdens on them: *physically*, in terms of the sheer expenditure of energy required to accomplish the day's work; *psychologically*, in terms of constant anxiety about solving the problems of daily subsistence, of social isolation, and of the almost continuous warfare in which women engage with the men in their lives; and *materially*, in terms of producing what the household needs to survive. These struggles, in a social environment in which virtually no "surplus" time exists for women – in poor households women tend to work three to four hours more than men (Cloud, 1985, p. 35) – and in a physical and technological environment that turns even the simplest tasks into a chore (one thinks of the hand-pounding of grain in rural communities), mean that women have typically lower educational attainments than men, are less free to leave the immediate vicinity of their homes, and live within a horizon of knowledge and interest circumscribed by their own children, their neighbors, and members of their extended family.

Women's social status does not reflect the objective importance that the performance of their domestic tasks would seem to warrant. A good deal of evidence shows that the increasing monetization of the domestic economy – its growing dependence on cash income in both rural and urban settings, but especially the latter – has reduced women's autonomy and strengthened male control over the household.[4] Although women have increasingly entered labor markets and are engaged in work that brings a monetary return, men still provide the bulk of household income.[5] Women's outside work is subject to frequent interruptions by

[4]   Addressing the question of whether women's condition has improved with the more complete integration of developing countries into the market system, Kate M. Young writes, "The general response is that women's condition has worsened; they are poorer, live in increasingly hazardous environments, and have lost the supportive mechanisms of the past. Such findings can appear somewhat contradictory given that studies on the impact of modernization suggest that in many countries general improvements in health, hygiene, housing, transport, etc., have until recently facilitated improvements in women's levels of health, education, life expectancy, etc." (1988, p. 2). She goes on to draw a distinction between women's condition and their structural position in society and asks whether "any serious improvement in their condition is possible without structural change." She concludes that "the structural issues are the central concern of the (less abundant) literature on women's position. This [literature] suggests that the social position of women, whatever their class, has worsened as a result of the integration of developing countries into the market, *regardless of whether or not women's condition has improved*" (1988, p. 3; my italics).
[5]   Conversely, women's unpaid/underpaid work represents a subsidy to capital, justifying wage levels that do not cover the subsistence needs of households.

pregnancies and emergencies at home, such as a child's illness, that may require their presence. Being on the whole less educated than men, women are also less skilled at many tasks. And many occupations, especially those with higher pay, are by custom reserved for men. Overall, therefore, women's earnings tend to be concentrated in the lowest-paid jobs. They also tend to earn wages that are only a fraction of men's. Even in Western countries, such as Italy, women earn less than men in *all* occupational groupings, although the ratio declines as women move from unskilled to skilled and professional work (Colombo et al., 1988).[6]

In the politics of households, women's access to and control of income turns out to be enormously important, providing a means for asserting their relative autonomy.[7] Household income is not necessarily pooled, and women earning their own money are frequently able to retain control, if not over all of their earnings, at least over a significant portion.[8] But the reason women venture into the market economy is generally not to enrich themselves. In virtually every case, mothering women will justify their employment outside the domestic economy by saying that it is to help feed, clothe and educate their children. They justify it by an ethics of care (Raczynski and Serrano, 1985, p. 249; see also Gilligan et al., 1988).

Both women's reproductive and nurturing roles and their limited but growing insertion into the market economy conclusively show that women's disempowerment is structurally determined. Women's agency – what they may and may not do – is severely restricted by their gendered identity, and their entitlements – from property shares to the food they eat – are similarly constrained. In both agency and entitlements, preference goes to the male.

And yet neither agency nor entitlements are fixed forever, although men are likely to cling tenaciously to power. In large matters as in small, both are always subject to renegotiation. It is this continuing struggle over agency and entitlements that gives households their political

[6] In Brazil, where women-headed households are said to constitute only one-fifth of urban households, 20 percent had no earned income at all, while women's participation in the labor force declined progressively from 60 percent at work yielding one-half the prevailing minimum wage to 7 percent at work yielding 20 times the minimum wage (Machado, 1987, pp. 56–60).
[7] As Raczynski and Serrano show, "The one who brings home the bacon also has authority, greater attributions, and greater freedom of action" (1985, p. 248).
[8] Women who are de facto heads of household will, of course, retain all of their own income. Children, who often begin work as soon as they reach five or six years of age, will be unpaid workers. Older children earning money may have to give up the bulk of their income. In households where men contribute financially, women may be responsible for only certain types of outlays, implying control, such as the education of their children. Unmarried young women working in assembly plants may be able to retain at least a portion of their earnings, while older women in informal work arrangements, such as selling home-cooked food on the street, are likely to retain all of their net earnings.

character. Although the politics of households may be relatively quiescent in rural communities, it becomes more agitated with urbanization, as traditional community controls over women's behavior become more relaxed and women begin to enter the labor force. Urbanization tends to unfreeze traditional patterns of social relations, with their disempowering effect on women, and sets in motion processes that eventually lead to women's emancipation.[9]

Women's claims

Women's claims within an alternative development fall into two categories. According to a widely accepted distinction first proposed by Maxine Molyneux (1985), claims may address either *strategic* or *practical* issues.[10] *Strategic* claims concern the systematic disempowerment of women that is encoded in social institutions. They address women's fundamental condition, proclaim women's rights, and seek to protect women against men's often aggressive and violent behavior. Strategic claims seek to change legal-institutional arrangements that keep women in a position of permanent subordination. It is therefore the fundamental claim of gender equality, and it is a long-term struggle.

If progress toward strategic aims must be measured in terms of decades if not generations, other, more practical claims require urgent attention and cannot be postponed. Indeed, making advances toward practical aims may also further the longer-term structural objectives. Development programs are never gender-neutral: the structure of opportunities available to women discriminates against them, and relative to men, women have substantially less access to the bases of social power and productive wealth.

An example of how development programs often deny women's participation is urban housing. Many women head their own households, yet they may lack the time or requisite skills to build housing or make home improvements themselves. Having no prospects of a steady income, women heads of households may not qualify for housing loans. In the countryside, women are often excluded from agricultural programs that

---

[9]   A large proportion of urban women, especially recent arrivals in the city, end up as domestic workers in middle- and upper-income households. In 1980s, for instance, 20 percent of the female labor force in the seven largest Colombian cities worked as live-in domestic workers (*puertas adentro*) and another 8 percent as daytime domestics (*puertas afuera*) (García Castro, 1989, p. 106). The "emancipation" of these women is something relative, as they find themselves in situation of virtual household slavery.

[10]   See also Young, 1988, and Moser, 1989. Whereas Molyneux (1985) chooses to speak of interests, Young finds it more useful to talk about practical *needs* and strategic *interests*. In the context of the present discussion, we will speak of both as *political claims*.

involve the introduction of cash crops or new technologies, even when they do most of the field work. In Bangladesh, for example,

> As men are usually the decision-makers and their activities are visible in rural societies, there is always a tendency for activities performed by them to be modernized in preference to women's. In most agricultural policies attention is focused on modernization and improved technologies for crop production which is done by men in the field rather than on post-harvest activities which is women's work. Moreover, development efforts to modernize or bring into the cash economy those subsistence level activities traditionally done by women, like rice processing, grain storage or poultry raising, tend to go to men and machines. As a result the expert knowledge of women in these areas is lost. (Abdullah, 1980, p. 36)

Women's structure of opportunities is also limited. Early pregnancies, undereducation, and domestic isolation all reflect the patriarchal regime, but the most pressing claims arise from women's restricted access to the bases of social power. All of the bases of social power mentioned in chapter 4 are gender-specific. We have seen this in women's differential access to housing and the restriction of their life space to the domestic sphere, the excessive amount of time women spend in subsistence activities, and their underdeveloped mastery of certain practical skills. Women's formal education in poor countries is nearly always lower than men's and of poorer quality. In Egypt, for example, women's illiteracy is reported to be over 70 percent (Ramzi et al., 1988, p. 199).

Information flows are similarly gender-biased. Women tend to know far less about the world than men do and generally rely on them to transact outside business, even when the outcome will directly affect their own and their children's well-being. Access to the tools and instruments of production is also severely restricted for women who lack property rights, have inadequate access to basic health services, and have very little control over their own bodies. Early and frequent pregnancies, high rates of self-induced abortion, hard work since early childhood, psychological harassment, poor nutrition, and polluted water all take their toll on women's health.

In all these concerns, as well as with income and credit, advancing women's claims would not seem to require an extensive preliminary phase of "raising consciousness." Women know very well what they need to start on the road to a better life. It is these very practical and immediate concerns that call for devising development programs with the needs of women in mind. The failure to do so – and the failure to address programmatic contents to specific forms of gender discrimination – not only will not help women to climb out of poverty but will jeopardize the attempt to incorporate disempowered households into the functioning economic and political spheres of their respective societies.

Gender and culture

Gender and gender roles are deeply embedded in a cultural matrix. It is the introjection of culturally learned attitudes and values that account for women's subordination and for the social reproduction of patriarchy from generation to generation, not in the least with women's own connivance (Papanek, 1990). Feminist rhetoric, with its high valuation of the autonomous, self-actualizing individual, is not pleasing to all women, especially when imported from the West. Throughout eastern Asia, for example, Confucian values stress social cohesiveness and harmony, and the individual's subordination is seen not to patriarchy as such but to the family. Personal meaning and fulfillment are found in filial loyalty and in shouldering such family responsibilities as custom decrees, with age being a major variable in addition to gender. Although individual hopes and attainments are not forgotten, they defer to the collective interests of the family and the larger community. What to Western eyes appears irrational conduct, given the value of "human flourishing", may seem to be perfectly in harmony with approved forms of behavior among Chinese families that satisfy their need for order and meaning in their lives.

An interesting confirmation of this cultural thesis comes from Carole H. Browner (1986) in a study of a Chinatec-Spanish-speaking rural township in the Sierra de Juárez in southern Mexico. Browner emphasizes the unusually integrated, resilient characteristics of this community of small-scale peasant producers, where men exercise exclusive political control and where they have successfully resisted repeated outside efforts to "modernize" the community's traditional way of life. The great majority of women express solidarity with the political decision to protect the closed, corporate character of the community. Their adaptive strategy to traditional subordination is to build strong relationships with their children, furthering their own goals by "manipulating interpersonal ties." As in the system of Confucian values, filial "piety" is expected: a mother's sacrifice for her children is expected to be reciprocated once the children have grown up. In Browner's interpretation, women's "way out" of a situation in which their own horizons are exceedingly limited and will remain so by political decisions made by the men in community is to concentrate on developing strong maternal bonds.

We may read this story as an instance of male hegemony standing against the kind of progress that might considerably ease women's burden. But in its original, Gramscian sense, hegemony means something more than simply domination. It is a form of domination whose legitimacy is ultimately based on a valued pattern of shared meanings. The women of the Sierra de Juárez are not straining to "liberate" themselves as individuals; their idea of "human flourishing" is very different from ours. Instead, they accept male domination in the public sphere as an integral part of a cultural pattern in which they gain strength from and personal satisfaction through their children.

This should not be taken to mean that gender equality in both its strategic and practical dimensions should not be put forward as a political claim. It does mean that culturally mediated meanings must be taken very seriously. Potential interventions in the household sphere go right to the heart of human relations. To assume that only technical or functional questions are at issue is wrong. Two Pakistani authors express this thought well:

> If the fledgling women's movement in Pakistan...ends up by operating outside the cultural parameters of its society it will face a problem of communication and identity. In order to communicate its message, the women's movement needs to use a language that is common to the people of Pakistan. Where language is a culturally defined system of signs and symbols, rejecting the cultural articulation of Pakistani society will impede the woman's movement. Secondly, recognizing that exponents of the patriarchal system label the women's movement "Westernized" only to discredit it, is not enough. Such a label can only be overcome if the women's movement is perceived by those it addresses being rooted in its own culture. At a different level, the efforts of Third World women to reconstruct their own women's history and to discover their own feminist legends reflects the same need to be rooted in their own history and culture. Therefore as long as Islam plays a central role in Pakistani culture, an Islamic framework is a necessity and not a choice. (Mumtaz and Shaheed, 1987, p. 158)

In other words, Pakistani women's struggle for liberation must be waged *from within* the national culture; it can neither be grafted on as an exotic import from the West nor propose the wholesale destruction of a way of life that has deep historical roots. If it is to involve the mass of women living in poverty, this struggle has to "make sense" to these women in terms of their own life experiences. To bring about a truly permanent change, the fabric of meanings must continually be rewoven.

Women's collective self-empowerment

The general approach to women's claims is one of empowerment. Empowerment has become something of a catchword among feminist writers in recent years, and different authors and practitioners have used it without bothering to attend to its different and shaded meanings. For some, it stands for social mobilization around women's major concerns, such as divorce, property rights, cost of living, peace, and the environment (Andreas, 1985). For others, it is a change in women's state of mind (Raczynski and Serrano, 1985; Logan, 1989). Here I have emphasized gains in access to the bases of social power. All three kinds of empowerment are in fact relevant to women's struggles. They may be thought of as forming an interconnected triad (figure 6.1). When this triad, centered on an individual woman and household, is linked up with others, the result is a social network of empowering relations that, because it is mutually reinforcing, has extraordinary potential for social change (figure 6.2).

Women working with other women on projects they find empowering – in production cooperatives, political movements, or mutual support groups – can accomplish a great deal more than a single woman acting for herself alone. Networking and organization – that is, acting collectively – tend to reinforce the process of women's social, psychological, and political empowerment. This is one of the lessons to be learned from social movements in Latin America, where social mobilization, often across class lines, has been a major force in bringing about adaptive and political change (Hardy, 1984, 1986; Feijoó and Gogna, 1985; Barrig, 1989; Jaquette, 1989; Nash, 1990).

Yet it is practical claims affecting livelihood that command the most attention of women in disempowered households. Four broad categories of need may be identified:

1 Time savings in the completion of household chores: solving the problems of potable water and fuel; acquiring improved cooking equipment, ready access to community facilities, day care for infants and small children, and better transportation to markets and services.
2 Improved health care, including birth control information and access to inexpensive prophylactic devices, abortion clinics staffed by professional women, infant care, and "barefoot doctors" who provide basic medical education and attention at neighborhood levels for village-bound women.
3 Acquisition of knowledge, skills, and information relevant to traditional women's tasks; for example, learning to read; learning how to improve personal hygiene, nutrition, and agricultural practices; mastering or upgrading artisanal skills; acquiring information about special services for women.
4 Expanded income opportunities from cash crops, small livestock,

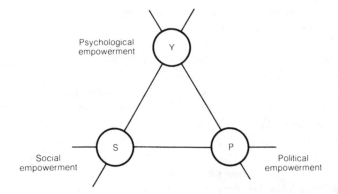

Figure 6.1 Forms of empowerment

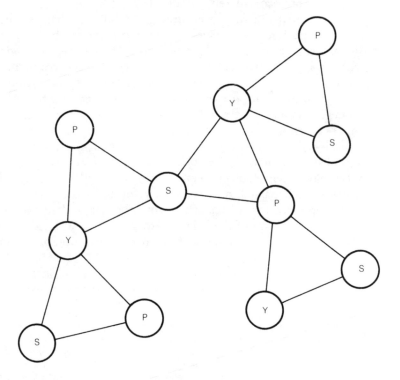

Figure 6.2 Empowering networks

artisanal production, factory work, and so forth; ensuring that women control their own earnings.[11]

With respect to these needs, women's political claims are for external assistance in a process of collective self-empowerment. This formulation emphasizes process, self-organization, social mobilization, and attention to the multiple claims of social, psychological, and political empowerment under culturally specific conditions.

When considered in the context of a specific national setting, these broad categories of claiming can be further specified and given substance, as this excerpt from an article on the condition of Sudenese women reveals: "The diversity of socioeconomic conditions across Sudan ... calls for a regionally differentiated policy.... For example, poor rural women need help most to raise food production ... to meet family nutritional needs and to market a surplus for additional income. Upperclass urban women most need consciousness-raising campaigns to redirect their resources to assist the poor in Sudan.... The present over-emphasis on skill training programs for which there is no demand in the market ... needs to be reconsidered" (Badri, 1990, p. 114).

I have stressed practical over strategic claims not because the former are more important but because response to them can be on a piecemeal basis, proceeding incrementally from village to village, neighborhood to neighborhood. Positive results in one place may be communicated to other places by word of mouth. Innovations are contagious. In any attempted solution, local women will take an active part.

Strategic claims, on the other hand, look to institutional and legal changes favoring women's status. Essential as they are for making progress locally, they are less likely to involve poor women, and their success in establishing a new framework for gender relations, even if formal changes have been made – a constitutional amendment adopted, a new law written and passed, a new ministry formed – will take a very long time. Compliance cannot be forced until there is widespread agreement on the basic premises of the emancipatory project. While deconstructing patriarchal power – beginning with the democratization of the household

---

[11] According to Kathleen Cloud, "In ... cases where women are not paid directly for their work, productivity often suffers, especially in systems with little pooled income. Women with high levels of responsibility for the provision of the family's basic needs and little access to cash are forced to substitute their labor, and that of the children they create, for the productivity-increasing inputs they cannot afford" (1985, p. 45). Cloud's own list of primary needs, at least for rural women, differs somewhat from the enumeration above. She identifies the following: (1) women's access to land; (2) women's access to capital, credit, and agricultural technologies; (3) women's access to household technology (e.g., cooperative grinding mills); (4) women's access to rural labor markets; (5) women's access to education and training (1985, pp. 39–45). An urban-oriented list of needs is developed by Amelia Fort (1988); for the special case of housing, see Moser and Peake (1987, p. 199).

– forms a very necessary part of the project, it is also by far the most daunting and difficult task, implying as it does a vast social and political transformation whose ultimate success will call upon enormous reserves of patience, determination, goodwill, and persistence. But the ultimate goal is very clear: because women's rights form part of a constellation of basic human rights, they must be inscribed in the very heart of an alternative development.

Futures Integration: Sustainability

Nineteenth-century Europeans and Americans were entranced with the idea of progress (Bury, 1920).[12] Bourgeois and Marxist alike celebrated what Manuel and Manuel in their magisterial study of utopian thought (1979) described as "an endlessly dynamic prospect founded upon the boundless expansion of science and technology, exploitation of the inexhaustible resources of the globe, and the flowering of human capacities" (cited in Martinez-Alier, 1987, p. 17).

The carnage of World War I put an abrupt end to this optimistic view of the world. A deep mood of despair followed when it became fashionable to talk about the decline of the West. The Great Depression, fascism, World War II, and the Holocaust provided ample evidence that, indeed, the barbarians were at the gate. This dark, foreboding mood was dispelled in the early fifties when Europe and Japan succeeded in rebuilding their devastated economies in record time, and when astounding rates of economic growth propelled the industrialized countries into an unprecedented era of affluence. Alongside these rekindled hopes, however, dissenting voices warned of impending doom. In 1972 the Club of Rome – a group of European multinational, managers, engineers, and scientists – published the epoch-making *The Limits to Growth*, which set off a chain reaction of studies to either refute or confirm its central message that a finite resource system, such as Earth, could not indefinitely sustain cumulative growth in population and production (Meadows et al., 1972).[13] Economics was not only about money; it had a material basis as

---

[12]   This section has greatly benefitted from the excellent review essay by Timothy Beatley (1989).

[13]   In an interesting commentary published 17 years later, Club of Rome President Alexander King claims that the message had been misunderstood. The problems identified in the original report, he said, although continuing to be of concern, should be considered "less as possible limits of economic growth than within the dynamics of the world problematique" (quoted in Pestel, 1989, p. 9). Overall, Pestel's revisiting of "Limits" reveals an unshaken belief in the powers of technology to *increase* resources ("technical progress will . . . lead to 'more from less,' i.e., less employment in manufacturing and less use of energy and traditional minerals resources") so that at least 50 percent of the world's population will be able to live more prosperously a half a century from now. Approvingly, Pestel quotes Club of Rome member Umberto Colombo to the effect that "man is now at the stage of being able literally to invent resources for his own use" (1989, p. 120).

well (Martinez-Alier, 1987). The globe's carrying capacity was not in-finitely elastic. In time, exhaustible resources would be exhausted, and chemical pollution would finish the rest. Great urgency invested the Club of Rome's dramatic message: apocalypse was less than a century away!

Whole new branches of knowledge – the environmental sciences – were suddenly flourishing, and a new discourse considered the possibili-ties of zero growth, steady state economics, and draconian measures (as in China) to scale back population growth to less than its replacement rate. The oil shock of 1983, African famines, the great dying of the forests in Western Europe, vanishing ecosystems in the tropics, the omi-nous hole in the polar ozone layer, the "greenhouse effect" threatening global warming, and other highly publicized disasters underlined the seriousness of the situation. It seemed as though industrial civilization – East *and* West – had reached a turning point. Unless changes were quickly introduced, the system would crash. But what new paths should humanity take? Some analysts (and most of the media) maintained their faith in salvation by science. Virtually unlimited energy sources could be tapped – nonbiotic methane gas deep in the bowls of the earth, nuclear fusion, the energy of the sun. Some even dreamed of colonizing space. Biotechnological inventions would solve the food problem. Other, equal-ly thoughtful, if less pollyannish, thinkers began to question the nature of a system that threatened humanity itself.[14]

What had gone wrong, and who was at fault? Was the systematic degradation and partial destruction of the material basis of human exist-ence – of life itself – a disease that afflicted only rich countries – their gout and obesity, so to speak – or were we all equally sick? There was reason to believe that the rich, industrialized countries were at the root of all the environmental evil. It was they who were using the bulk of the world's commercial energy resources, they who were the most wasteful nations, with Canada and the United States in the lead (table 6.1). Relative to its population, Japan consumed less than one-half of Amer-ica's energy, while poor countries made do with a great deal less.

Table 6.1 suggests two additional conclusions. The first is that most countries, with the United States again in the lead, are having to devote significant and apparently growing proportions of their exports (which themselves are increasing) to the importation of petroleum and other fuels. By 1987 non–oil-producing countries were spending up to one-fifth of their exports for needed energy. The second conclusion is that con-tinued economic growth appears to be closely correlated with energy

[14]   Michael E. Colby (1990) has contrived a typology of environmental ideologies that is helpful in untangling the strands of environmental discourse. He has distinguished five "paradigms": frontier economics, environmental protection, resource management, eco-development, deep-ecology.

Table 6.1 Commercial energy consumption, 1965–87

	Per-capita energy consumption[a]		Percent change	Percent (US = 100)	Energy imports as percent of exports	
	1965	1987			1965	1987
China	178	525	195	7	0	2
India	100	208	108	3	8	17
Other low-income countries	73	116	59	2	8	16
Lower-middle-income countries	531	863	62	12	8	10
Upper-middle-income countries	653	1,392	113	19	8	12
OECD members	3,748	6,573	75	90	11	12
Other high-income countries	1,943	3,030	56	42	7	7
Nonreporting countries	2,509	4,777	90	66	–	–
Japan	1,474	3,232	119	44	19	17
United States	6,535	7,265	11	100	8	19
Canada	6,007	9,156	52	126	8	5

a Kgs of oil equivalent
Source: World Bank, 1989a, table 5.

demand: between 1965 and 1987 per-capita commercial energy consumption in all countries except the United States increased by one and a
half to three times the original base. Mainstream development in the
second half of the twentieth century was clearly energy-intensive.[15]

But if rich countries are characterized by energy wastefulness and a
growth pattern that is clearly unsustainable in the long run, poor countries are apt to impair the physical basis of their economies out of sheer
necessity. It is necessity that has stripped some of the steepest mountainsides in Haiti of their forest cover, while peasants, suspended from ropes,
plant, mulch, and harvest maize in otherwise inaccessible locations. It is
necessity that has accelerated the crop-rotation cycle as population press-
es against arable land in southern Nigeria. It is dire necessity that causes
Indian peasants to burn as fuel the cow dung with which they could
otherwise fertilize their fields. And it is economic pressure that forces
poor countries to produce crops for export to European and American
markets while neglecting the production of basic cereals for domestic
consumption.

Although poor countries are not immune to environmental disasters,
the elementary division between rich and poor is in many ways too crude
for the answers we seek. We have to look more closely at the distribution
of the environmental burdens of "development." It is the relatively rich,
especially those living in large cities, who most use and pollute a poor
country's resources. But the rich are only dimly aware of the damage they
cause. They are able to buy themselves relatively unpolluted enclaves,
surrounded by spacious gardens, while the least attractive, rankest neighborhoods are left to the poor. Environmental pollution is most visible
(and least bearable) in the squalid city quarters where over half the
population raise their families – ramshackle houses extending precariously over open sewers, subject to periodic flooding and overrun by vermin.
It is a familiar story that issues in periodic attempts at housing reform,
police repression, meliorative action, and occasional outbursts of popular
wrath (*Environment and Urbanization*, 1989).[16]

---

[15]   The observed slow-down in energy consumption in the United States was partly a result of
deindustrialization and the shift of production to the service sectors. However, the country
imported huge amounts of "embodied" energy in the form of finished commodities produced
abroad, especially Japan.

[16]   For a neo-Darwinian evolutionary interpretation of economic growth, see Adams, 1988.
According to the author, who is an anthropologist, the close association between gains in
productivity and increases in commercial energy strengthens what he calls the regulatory classes
of society, or elites, while marginalizing others. Adams concludes by listing five sets of evolutionary consequences of the worldwide increase in production in the decades following World War II:
population explosion, reduction of ecological and species variety, reduction of cultural variability
at the world level, increased level of living differentiation, and increased energy flows and
complexity (1988, 238–9). His theoretical framework, reminiscent of a long-discarded social
Darwinism, nevertheless has the advantage of offering a systemic interpretation of phenomena
that are usually treated in a fragmentary manner.

The rich in poor countries imitate the consumption patterns of the rich in rich countries, supporting a type of economic growth modeled on the energy-intensive pattern of the world metropolis. In short, those who profit most from the system in dominance are also the most likely to ignore Herman E. Daly's "impossibility theorem," which "simply states that a U.S.-style high-resource consumption standard for a world of 4 billion people is impossible. [World population is projected to reach more than 6 billion by the year 2000 and more than 8 billion by mid century.] Even if by some miracle it could be attained, it would be very short-lived.... Crises of depletion, pollution, ecological breakdown would be the immediate consequences of generalizing U.S. resource consumption standards to the whole world" (1980, p. 361).[17]

If Daly's impossibility theorem turns out to be true – and there are reasons to believe this the case – imitative development by and for the rich in poor countries may be called a form of *misdevelopment*, a process that shifts the real social and environmental costs onto the disempowered sectors of the population. By the same token, rich countries have rolled off the environmental costs of their ecologically unsustainable *overdevelopment* onto poorer (disempowered) countries, repeating the strategy of attempted insulation against the full consequences of the growth pattern rich countries have chosen (Netherlands Ministry of Housing, 1989, chapter 2). Rich countries export polluting industries to newly industrializing ones. Rich countries promote an energy-intensive, export-oriented agriculture in poor countries that not only exceeds their own financial means but makes them dependent on grain imports from the United States and Canada. Rich countries try to dump their toxic wastes in countries on the world periphery. And rich countries, after having pushed poor countries into high levels of indebtedness, urge them to adopt economic policies and institutional reforms (free trade competition, export-orientation, privatization) that take them ever farther from sustainability along the tragic path of misdevelopment.

But the environmental crisis has also made rich countries, especially in Western Europe, more sensitive to some of the transnational consequences of *overdevelopment* such as acid rain, reduction in species diversity, global warming, oil spills, and nuclear accidents. Among other responses, these countries now insist that poor countries adopt resource-conserving strategies. Tropical forests should be preserved, big hydroelectric dams should be abandoned in favor of small watershed programs, and so on (Schwartzman, 1986; *Financing Ecological Destruction*, 1987). It is less clear that rich countries, which constitute a shrinking

---

[17]   Daly currently serves as senior adviser to the World Bank on environmental questions. It is interesting to compare his dark pronouncement with the more upbeat assessment by Pestel (1989), who envisions the progressive march of industrialism, with technology in the lead.

proportion of the world's population but consume the bulk of its resources, are prepared to adopt resource-conserving principles and change
their own consumption patterns. Some years ago, a Swedish proposal
along these lines was met with scorn and ridicule (Lindholm, 1976).

The case for intergenerational equity

"Sustainability," said the Brundtland Report, is a development "that
meets the needs of the present without compromising the ability of future
generations to meet their own needs" (World Commission, 1987, p. 43).
This attempt at a concise definition reveals sustainability to be a slippery,
ambiguous concept. Let us try for a clearer understanding by translating
the World Commission's concern into our own language.

Of all the disempowered groups within a population, those with the
least power lack an effective voice in public affairs, above all infants and
children, who are virtual citizens of their respective countries, and the
generations of the unborn. If, on the other hand, an alternative development is principally about empowerment, then all those who will inherit
the world we have made must come into the public forum with us and
state their claims. And their central claim is the right to inherit from us
an environment in a condition at least no worse than that in which we
received it. Their claim is for intergenerational equity or fairness in the
distribution of environmental costs and benefits.

Unlike Schumpeterian capitalism, which in its forward surge of "creative destruction" never looks back, an alternative development respects
the traditions of territorial communities. It respects that their historical
continuity is also a major source of personal meaning and collective
identity. As members of specific communities, we form bonds not only
with the living but also, and this is the critical point, with the generations
that precede and succeed us.

It would be absurd to argue that human bonds arise exclusively from
territorial relations. People nurture solidarities for many reasons: kinship,
religion, and political ideology are but a few. But territorial clusters of
human settlements do bind us, not the least culturally and politically. We
may choose to leave our birthplace or we may be forced into exile. In
either case, we only change places. Sooner or later we will want to claim
citizen rights in another community and accept another history as our
own.

Political communities exist in time as well as space. It is their *continuity* that we enshrine in the collective memory we call history and that
makes us look at our children and their offspring with joyful anticipa-
tion. Historical continuity assumes the existence of a pangenerational
human bond.

Whether as virtual or as active citizens, we demand accountability of state and corporation, while of each other we expect social behavior that is measured by the norms of good citizenship. By extension, future generations can demand public accountability of us. We the living are no more than the world's transitory custodians. The question is, How good is our stewardship?

As we search our collective conscience for an answer, a central issue concerns the number of generations we are prepared to admit into the forum. What are the temporal limits of our accountability? In democracies, politicians tend to think only of the next election, which may be a few weeks away. Corporations are guided by the opportunity costs of capital and the rapid rate of technical and product innovation. In politics no less than in the corporate economy, the short run is what matters. Behavior in these domains appears to abide by the cynical dictum of Louis XV, *"après moi le déluge."*

At the other end of the spectrum are the voices of ecological fundamentalists. For them the problem with nature is the human race, and the best of all possible worlds would be a global wilderness. In environmental matters, fundamentalist thinking is tuned to a cosmic scale (Devall and Sessions, 1985).

Although the short run is the optimal horizon for speculative capital and a democratic politics, it is precisely for what happens in the short run that future generations will hold us answerable. We are not free to use up the world's resources in our own lifetimes, practicing, a kind of scorched-earth policy out of sheer self-indulgence or carelessness. But over how many generations can we be held responsible?

Airports, railroads, dams, tunnels, bridges, electric power installations, and even housing have useful lifetimes of up to 100 years. The human lifespan is between 70 and 80 years (in poor countries it is 10 to 20 years less). A grandchild born to me today can expect to live well past the middle of the twentifirst century. A relevant time horizon of three generations, or 75 years, does not, in this view, seem extravagant. Our forward look with regard to environmental consequences must be on that order, and social discount rates used for analyzing public investments should reflect pay-out periods of at least three generations and consequently run considerably *below* current market rates with their high time preference.

Unfortunately, when we make decisions for ourselves as individuals we tend to avoid a social perspective. We may rationalize this by saying that, unless they are closely watched, our fellow citizens will likewise fail to live in socially responsible ways. We may argue that no one in fact can foretell the changes in knowledge, technology, relative prices, and future demand that are likely to occur over the next three generations. But experience has shown that when individuals act exclusively in their own self-interest and outside the sphere of social control, common resources will be used up in excess of what the community would have decided had it convened to debate the matter.

The power elites can escape at least some of the consequences of this well-known "tragedy of the commons" by spending a lot of money on themselves. Behind security gates, they are able to create Edenic environments. But the vast majority of the people do not have that option.

Because the historical future hides behind a veil of time, the way of dealing responsibly with radical uncertainty and high risk is not to harbor a blind faith in science and technology, and therefore hope to minimize risk, but to open up a political discourse and begin assessing the real risks involved. Both the survival of treasured lifeways and escape from poverty require collective action within an inclusive democratic order. Risks and consequences must be equitably shouldered rather than rolled off onto future generations. We need, then, a discourse that asks the forbidden question about the limits to growth. For this, we need more than purely "rational" criteria. We must engage people's feelings as well; the discourse must take an ethical turn. In the last analysis, sustainability poses a question concerning the nature of the good society.

Implementing sustainability

Political claims are empty if they cannot be linked to appropriate means for their realization. What, then, is required to introduce an explicit future dimension to an alternative development? I approach this difficult subject by looking briefly at four examples of doing things right.

1. *Getting the prices right.* One of the most effective ways to ensure intergenerational equity is to internalize the social and environmental costs of production in the price structure of commodities. This can be done in two ways: by obliging producers to make the investments necessary to decrease waste generation (e.g., recycling water, reducting at the source, building double-bottomed oil tankers) and by imposing taxes either on producers or final commodities that reflect society's valuation of the environmental costs incurred in their production (e.g., resource depletion, costs of recovering degraded or lost environmental functions). The effects of incorporating social and environmental costs into the structure of prices will be to *reduce* the demand for products that are highly damaging to the environment and to *shift* demand to less aggressive sectors of the economy.[18]

2. *Environmental accounting and research.* No matter how accurate-

---

[18] Possible regressive effects can be relieved with consumer subsidies.

ly markets reflect the "true" costs of production by taking third-party effects into account, they remain an insufficient guide to intergenerational equity. In addition to prices, we need a type of accounting that will describe in physical rather than monetary terms exactly what is happening to the resources of air, water, land, forests, biological species, and strategic minerals (Hueting, 1980). Taxing ivory will not stop the slaughter of elephants, nor will increasing the cost of tuna diminish the dolphin victims of monofilament nets. No one can impose fines on the famished peasants of Haiti when they cultivate row crops on steep mountainsides unsuited to anything other than trees or grass. And the construction of big hydroelectric dams that threaten to flood enormous acreage of rain forest and destroy the habitat of forestdwellers is not price-responsive. Material, measures are needed to gauge the effects of economic growth on the environment and to devise countervailing strategies, including further research, outright prohibition, new engineering designs, and appropriate regulatory measures.

3. *Achieving food security.* Like so many terms in the policy sciences, the meaning of food security has rarely been spelled out, although policies looking toward food security are frequently advocated. One of the best accounts of what a system offering food security would mean has been sketched out by the Chilean agronomist Alejandro Schejtman (1983):

● SUFFICIENCY, meaning the system's capacity to generate a sufficient internal food supply (via national production or imports) to meet expanding demand and the basic food needs of all.

● RELIABILITY, signifying a control of food system mechanisms, so that seasonal and cyclical production fluctuations, especially of staple foods, are minimized. Increasing reliability would include lessening the importance of seasonal (nonirrigated) crops, eliminating the often erratic effects of food subsidies, and improving transportation, storage, distribution, and so forth.

● AUTONOMY or SELF-DETERMINATION of a food system, requiring maximum reduction of its vulnerability to the uncertainties of the international marketplace (hence rendering it more dependable). But autonomy signifies neither autarchy nor self-sufficiency. On the contrary, specialization and regional trade should be stimulated. The preconditions for specialization and comparative advantage, however, cannot be dictated solely by market considerations.

● LONG-TERM STABILITY, meaning a food system's ability to provide sufficiency, reliability, and autonomy without destroying the base of agriculture – the ecosystem. Agricultural modernization (with the United States serving as the model) is, in terms of long-term stability, a chimera. It has been estimated that, in the United States, more than over nine calories of fossil energy are needed to produce one calorie of food. Although the corresponding ratio for France, estimated at 5.5 to 1, is lower, world fossil energy resources would not permit widespread imitation of either model.

● EQUITY, which is the final and perhaps most important attribute of a food system offering acceptable levels of nutritional security. Although difficult to define, it involves, as a minimum, access to sufficient food by all social groups and individuals. This implies that the food system is shaped in part by the interests of the majority. In poor countries it also means curbing the luxury consumption that could sabotage meeting the first four goals.

Even a cursory review of these characteristics shows that a food security system in this sense would significantly contribute not only to sustainability but also to other dimensions of an alternative development, such as improving peasants' productivity through low-input agriculture and upgrading indigenous agricultural production systems.

Throughout the world, many sustainable development projects are in place that conform to Schejtman's criteria: windbreaks in central Niger, watershed restoration in the Himalayan foothills, rural development in the Dominican Republic, and cassava and integrated pest management in a number of African countries (Reid et al., 1988). These projects and policies seek an upgrading of existing production systems rather than a wholesale replacement by less equitable, more energy-intensive and fragile production systems geared to compete in export markets. They take an ecological and culturally sensitive approach to increasing productivity.

4. *Building energy-conserving cities.* A widely held view has it that the environmental problems with the most direct bearing on sustainability are "out there" in nature – the forests and fields, the streams and lakes of the unbuilt environment. But the true cause of rural degradation is urban-industrial civilization. It is the imperial city that reaches far beyond its borders to feed itself and to satisfy its hunger for energy, spewing out what it cannot digest as poisonous effluent. It is the imperial city that forms the countryside in its image.

Strangely enough, the question of making cities more energy-conserving has not yet captured the imagination of many people. A few architects have worked on passive energy designs, but the major statement pertaining to Third World cities is a little-known monograph, now ten years old, on resource-conserving urbanism (Meier et al., 1981). Richard L. Meier's study is suggestive rather than definitive; it lacks quantification and has no theoretical pretensions. Nevertheless, it is full of astute observations concerning particular cities, such as metropolitan Manila, and replete with practical proposals for making the Third World megacity environmentally more benign. The study's many ideas include the following:

- Irrigated urban fringe for market produce. Six to ten crops per year. Accompanied by small livestock development (chicken, rabbits, ducks).
- Complex of tall trees, fruit trees, and tubers around human dwellings that are built at two to three times village densities.
- Adaptation of traditional building forms.
- Principal road network, surfaced for low-friction traffic, maintains continuous bus traffic; side roads encourage three-wheelers, bicycles, and carts.
- Multipurpose workshop and office buildings that maximize ventilation internally, replacing air-conditioning.
- Aquaculture of green vegetables, carp, tilapia, and fully domesticated milkfish.
- Chlorella ponds for sanitation in many low-lying areas.
- Floating suburbs of independent houseboats, serviced by duty ships delivering food, fuel, water, and mail and collecting wastes and home manufactures for disposal.
- Integrated, solar-energized neighborhood water centers that provide water for baths, food preparation, laundry, artisanry, toilets, and waste removal.
- Use of communications satellite for reducing air trips to provincial cities by providing a series of telecommunications and computing services, including advanced telepostal services to outlying points. Equipment repair, medical advice, education, commerce, and consulting of government officials can all be expedited. (1981, pp. v, 139–40)[19]

How feasible some of these ideas will be once they are tested in specific urban settings remains to be seen. What is significant is the authors' general approach to energy-conserving urban design: searching for a symbiotic relation between "natural" processes (social forestry, aquaculture, chlorella ponds) and urban functions, substituting solar energy for fossil fuels, substituting low-energy communications for high-energy travel, adapting traditional architectural forms to contemporary uses, and recycling wastes. The authors conclude that by avoiding imitative

[19] The report borrows freely from an earlier study by Meier (1980) with application to Jakarta. The following proposals may apply to either Asian metropolis.

design solutions, overall energy consumption in a city like Manila could be reduced by as much as 50 percent.

Studies such as Meier's could be carried out systematically for a number of Third World megacities, and pilot tests could be undertaken to verify many of its challenging proposals. When done, these tests are likely to show that imitative urban development leads indeed to *mis*development, and that practical means are at hand to ensure that even cities on the world periphery can do their part in becoming environmentally sustainable.

The four methods of reaching intergenerational equity or sustainability we have discussed − right prices, environmental accounting and research, achieving food security, and building resource-conserving cities − are complex. They look toward keeping the environment intact despite population increases and continued economic growth. But this simple homeostatic objective hides from us the larger implications of an alternative development − something that points to more than lower discount rates, alternative technology, and sewage recycling plants. It suggests a particular way of thinking about the relations between humans and energy systems and among humans themselves. Underlying an alternative development is the assumption that humanity is not locked into immutable evolutionary laws (Adams, 1988) but that, within limits that must be respected, we have the ability to achieve a socially just, sustainable, and satisfying life for everyone.

Advocacy and environmental action

Policies such as maintaining food security and making cities resource-conserving are not likely to be adopted unless a country's government is pushed to do so. As always, politics is at the center of the action. And politically inconvenient policies will not be adopted unless countervailing power can be mobilized. Mobilized citizen power culminates in environmental actions that are typically in opposition to misguided development initiatives at the local or regional scale. Consider the following:

- Brazilian rubber tappers and native forest dwellers in the Amazon are mobilizing against repeated, large-scale incursions of habitat by cattle ranchers and other commercial interest. They invented "extractive reserves" as a means of protecting their livelihood. (Hecht and Cockburn, 1989)
- Movements of small peasant farmers in southern Brazil directed against the construction of large-scale hydroelectric dams that would flood their farms and deprive them of their traditional means of livelihood. (McDonald, 1989)
- The peasant women in the lower Himalayas who depend on the

forest for wood for fuel and building, grass, vegetables, honey, medicinal herbs, and fruits and who literally hugged their trees to prevent them from being cut down by commercial loggers. Cooperative afforestation efforts have resulted from these early actions. (Bhatt, 1990)

- The movement to develop an ecologically sound habitat by squatters in Mexico City's green-belt zone of Ajusco was originated in response to repeated government efforts to dislodge them from the site. (Pezzoli, 1990)

Struggles such as these are not only empowering but frequently serendipitous as well, leading to new environmental practices. But to have a lasting effect, these social inventions need to be raised above the specific local instance. Extractive reserves, appropriate technology for urban settlements, more efficient energy use that would scale back needed power-generating capacity, and social (village) forestry in the management of fragile hillside environments, all of which began as "spontaneous" answers to specific livelihood issues, have evolved into major policies promoting intergenerational equity in resource use. Environmental action is part of poor people's struggle for survival (Agarwal et al., 1987), but unless it leads to more encompassing policies (and practices), it will not lead to an alternative development.

Concluding observations

The environmental question to which the formula of intergenerational equity refers is not a one-way street, applicable only to poor countries. Rich and poor constitute a single world system, and the overdevelopment of the first is closely linked to the misdevelopment of the second. Neither "development" is sustainable in the long run; both fail to meet the equity test. A version of alternative development is thus as pertinent for the countries central to the world economy as it is for those on the periphery.

In an early exercise of "green thinking," Nordal Åckerman, a professor of history and national security, asked whether the Swedish standard of living could be maintained in the face of global "limits" (1979). It is a question that must be asked with growing insistence of the citizens of all industrial and "postindustrial" countries.[20] Our present lifeways are both

---

[20]   The language of postindustrialism has become quite popular in rich countries, but the term is misleading. Although it is true that a progressively smaller percentage of the work force in these countries is engaged in manufacturing, we continue to consume manufactured products at an increasing rate. The difference is that these goods frequently originate in poorer, industrializing countries (such as Mexico and Korea) that have become integrated into the global economy, even as manufacturing processes within rich countries are becoming more capital-intensive, employing less labor per unit of value added. The perception of a postindustrial society is superficially linked to domestic employment in rich countries. A global perspective is needed to capture the actual trends.

inequitable and ecologically inappropriate. They are not sustainable. We delude ourselves that there is nothing to worry about, because we can turn on the heating system in winter and air-conditioning in summer, and because petrol can still be had (in the United States) for little more than a dollar a gallon, because our garbage is punctually picked up each week and carried away (we know not where), because the supermarkets are well-stocked, and because we can still escape into a relatively unspoiled countryside on weekends. We may believe that, for example, drought conditions and sandstorms in Botswana have nothing to do with our way of life. But they do.

The degradation of Botswana's rangelands, caused by overgrazing, has been encouraged by a succession of World Bank loans that were intended to boost beef exports from this southern African country to Western Europe. But overgrazing was not the only destructive consequence of this misguided policy:

> The cattle industry in Botswana has taken a disastrous toll on Botswana's once plentiful wildlife. As a result of EEC requirements that Botswana beef be free of hoof-and-mouth disease thought to be transmitted from wildlife, the country has erected more than 800 miles of "veterinary cordon" fencing designed to separate wild animals from cattle. Fencing has caused the deaths of hundreds of thousands of antelopes and other wild animals in recent years. The effects of massive numbers of cattle also threaten the traditional way of life of the Bushmen hunter gatherers who depend on the undisturbed environment and wild game for survival. (*Financing Ecological Destruction*, 1987, p. 8)[21]

To turn Botswana into an African Texas is thus the wrong kind of help. But to recenter its economy (and that of other beef-exporting countries, such as Brazil) on a rural development based on the principles of food security might ultimately require that the meat-eating countries of Western Europe adjust their own consumption habits. The environmental problems of Botswana (and Brazil) have become our problem.

The unsustainability of lifeways in the world's rich countries is gradually beginning to be recognized. We are shocked by the spread of violence, drugs, and pollution in urban America. The steady upward climb of per-capita income is no longer viewed as an indication that the quality of life is improving. It has cost more than $50 billion to house America's half-million prisoners. Will building more jails to house the steadily growing population convicted of criminal acts – jails whose construction cost is shown as a *gain* in national income – improve the

---

[21] For an interesting historical and class analysis of Botswana's "enclosure movement," see Worby, 1988.

quality of life for the average American? We are deluded into thinking that the way we have been doing things is either socially or environmentally sustainable. It is not only Sweden's economy that needs to be shrunk and made over. A major rethinking of where America is going is also in order.

Alternative Claims: A Review

As emphasized at the beginning of chapter 5, the claims being made for an alternative development correspond to different problems of inclusion: political, economic, and social, with a fourth claim reaching out to include future generations that have a stake in the present. A comprehensive claiming is intended.

The normative bases of these claims, argued in chapter 1, are the rights we enjoy both as human beings and as citizens of territory-based communities. I would insist on the territoriality of an alternative development for a number of reasons:

- Territory is coincident with life space, and most people seek to exercise a degree of autonomous control over these spaces.
- Territoriality exists at all scales, from the smallest to the largest, and we are simultaneously citizens of several territorial communities at different scales: our loyalties are always divided.
- Territoriality is one of the important sources of human bonding: it creates a commonweal, linking present to the past as a fund of common memories (history) and to the future as common destiny.
- Territoriality nurtures an ethics of care and concern for our fellow citizens and for the environment we share with them.

The political claims of an alternative development originated in organized civil society. Masses of people – the urban popular sectors, the small-farm peasantry – are only marginally incorporated into their societies: economically they are redundant, politically they are excluded. By claiming the right to full inclusion in public life, they seek empowerment for themselves. But the empowerment they claim is not to seize the state. Their demand is for social justice and a respectful treatment as citizens with equal rights. In the ecology of power, it might appear as if strengthening the disempowered sectors of civil society would weaken the state, as if society were some kind of complicated zero-sum game. But this would be to draw the wrong conclusion. Although an alternative development argues for a broad devolution of powers to local and regional levels of governance, a restructuring of the areal division of powers is likely to strengthen the central state in three ways: (a) by disengaging the state from problems that are better dealt with at local and regional levels; (b) by creating institutions that can be responsive to the diversity of locally and regionally articulated needs; and (c) by stabilizing the political system. Strong states tend to have strong civil societies. And where civil society is strong, we are likely to find a complex areal division of powers, with local governments playing a major role.

Our discussion of alternative claims identified several gaps in knowledge that can only be closed with more research: a thoroughgoing revision of national product accounts to net out major social and environmental costs; systems of social indicators, particularly at disaggregated levels; resource accounting in physical terms; and experiments with resource-conserving urbanism. This research will undoubtedly be expensive, although not excessively so; it is essential for making political assessments of the quality of economic growth.

Qualitative growth occurs whenever better is judged to be better, not when it is simply more. And to judge the quality of a complex phenomenon such as economic growth requires something other than a single criterion. It needs a composite measure that involves attaching different weights to its components. By analogy, music competitions are run with an invited jury of experts to judge the performances according to standard criteria such as musicality, interpretation, technique, memory, intonation, poise, and so forth. Each judge assigns a numerical score to each criterion, which enters the final score with a predetermined weight. The verdict is based on an averaging of all these scores. In the case of economic growth, quality is similarly determined, only the jury members are now citizens, and the weights grow out of the constituted political process.

The level of aggregation is critical here. The more local the assessment of quality, the greater will be the territorial diversity of what is thought to be "better." In a world that is in constant flux, diversity is preferable to high levels of territorial aggregation. This is as true for countries, where regions and cities must be broken out as significant territorial entities, as it is for individual cities, where it is districts and neighborhoods that are important, each requiring its own fine tuning. Mainstream development tends to wipe out cultural and ecological diversity. An alternative development seeks to restore and preserve it.

If measures of quality must be stitched together in different ways, as if the object were to make a great variety of multicolored quilts, one for each city and rural district then, at more aggregate levels of territoriality quality must be understood as a colorful patchwork made from all these quilts, although an increase in scale may require additional thread work.

Questions of distribution and social justice, for example, are best handled as an aspect of quality at larger territorial levels.

Finally, the most fundamental requirement of an alternative development is an inclusive democracy along with all the institutions that pertain to it. Without democracy, there can be no politics, and without a genuine and inclusive politics, the claims of the disempowered will not be heard. In some countries heads of local government are still appointees of the minister of the interior, who is also the minister responsible for the national police. In the absence of an inclusive democracy, local government becomes an agency of repression.

We are living through a period that, for many parts of the world, has ushered in democracy and dismantled totalitarian and authoritarian regimes. Spearheaded by social movements in Latin America, Eastern Europe, and even Africa, it holds out an exciting prospect. Yet it is far from clear what kind of democracy will be found at the end of the rainbow when a system has stabilized. Will it be a democracy only in form, or will it also have the pungent flavor of everyday life? Will the newly wrought system tap into the strong democratic talk of civil society? Will it accommodate dissidence? And above all, will it be a democracy that resolutely undertakes the great task of social, economic, and political inclusion?

Unable to look over the rim of history, we have no answers to these questions. But the stress on local action should not blind us to the need for structural reforms. As people begin to acquire social power, they must not neglect the political power necessary to advance their interests and concerns. The state is unlikely to act on its own accord to eliminate the structural constraints that serve to exclude people from effective participation. In the next and final chapter I address this problem of translating social power into political power. Without a politics of social movements, and without a political system capable and willing to respond positively to citizen claims, an alternative development will be stillborn.

7   Practice: From Social Power to Political Power

This chapter is about implementing an alternative development. I focus on Latin America because my experience there is the most extensive. The political context, therefore, is that of the democratic transition since the mid 1980s (Lehmann, 1990), as well as the debt-ridden nature of Latin American economies, their lack of dynamic growth, and the rise of civil society. Above all, the context is the practical utopia of political and economic inclusion to which the following excerpts from some of Latin America's leading intellectuals give testimony.

The urban planner Alfredo Rodriguez wrote the following in 1983, ten years into Chile's military dictatorship:

> To think of a democratic city is to think of a mode of inhabiting, participating in, building, creating, living, and imagining the city, the urban stage, new forms of space.
>
> To think of a democratic city, at least as possibility, means to do so along different dimensions which express the new relations that will exist among its inhabitants, the citizens.
>
> How shall the city be governed?
> How shall the citizens participate?
> How shall public services be decentralized in democratic fashion?
> How shall urban spaces be appropriated collectively?
> How shall the city's public spaces be structured?
> How shall relations among citizens be expressed through urban space?
>
> Any formulation is restricted and partial to the extent that it limits the project of democracy at the level of the city to universal suffrage, the secret ballot, and the election of public officials. The debate about urban democracy must include the possibility of new forms of generating and appropriating urban space, so that its use value may be preserved.
>
> It is necessary to create new options for everyday life and to open the door to the emergence of new architectural forms and urban designs that grow out of, even as they generate, new forms of the convivial life. (1983, p. 47; my translation)

At a time when one-third of Chile's labor force was out of work, and when the democratic resistance to the military dictatorship was gathering force, these were visionary words. Another six years would pass before General Augusto Pinochet was defeated at the polls, and a period of democratic transition initiated.

Rodriguez affirms his faith in an inclusive democracy. He addresses the people of Santiago, and indeed all Chileans, in their roles as citizens. He describes representative democracy as partial and segmented, envisioning a "strong" democracy capable of penetrating deep into civil society and everyday life. And he expresses a powerful desire to reclaim the city from the hands of real estate speculators and to restore its "use value" as generator of a public – that is, of a civic and convivial life.

Rodriguez speaks with the voice of a civil society that was beginning to reclaim what it had lost after ten years of enforced silence. He echoes a sentiment that is now general throughout Latin America and many other parts of the world, including Eastern Europe and the Soviet Union. As the twentieth century draws to a close, civil society enters the public domain as a collective actor on its own behalf, calling for a new and inclusive democratic politics.

The Peruvian sociologist José Matos Mar speaks of a *desborde* (flooding) of the traditional channels of authority and governance by the millions of Indians who have come down from the sierra (the Andean range) to live in Lima. Serrano popular culture, says Matos Mar, is beginning to change the traditional rules of the game by which politics in Peru is conducted and to forge a new, more democratic and inclusive society (1985).

The Argentinian José Luis Coraggio, now living in Quito, Ecuador, writes of Latin America's "search for new utopias" and the growing presence of what he calls *lo local* in the reconstruction of society (1988). The state has exhausted its possibilities for development, Coraggio asserts. Civil society is the new fountain of energy. Coraggio's vision is of a popular, participatory democracy linked with a strong and socially progressive central state. In a rediscovery of Montesquieu and Rousseau, he sees a territorially organized, autonomous civil society as the true foundation of a reconstituted state.

Aníbal Quijano, a Peruvian *pensador* (intellectual) who spent years working in the Social Affairs Division of the Economic Commission for Latin America, writes passionately about "another modernity" based on culturally endogenous forms of life in the Andean region that stress individual liberty, reciprocity, everyday democracy, equity, and solidarity among equals (1988). He sees this "other" modernity giving rise to a new sense of what is private: this sense he calls *lo privado social*, which, together with its nongovernmental institutions, ultimately gives form not only to a new kind of civil order but also to a new kind of state.

Pedro Santana, a Colombian political scientist, studies the *paros cívicos* (civic strikes) that have become a popular means of social mobilization in his country. Between 1971 and 1980, 128 such strikes were counted in Colombia, mostly in small and secondary cities, but also in a few regions, and at least two on a nationwide basis, involving millions of people protesting government inaction or misguided action. In Colombia, a *paro cívico* is a strike that leads to the complete paralyzation of productive, commercial, educational, and administrative activities. It is a broadly based social movement, spontaneous in that it is not organized by any political party or *sindicato* and constituting an impressive lesson in unmediated civic politics. "Notably," writes Santana, "over the course of the movement, more often than not, social cohesion breaks through the framework of collective demand-making to reveal an ample practice of direct, participatory democracy in decisions concerning not only the just demands to be pursued, but also the forms of the struggle to be adopted" (1983, p. 113).

Colombia's *paros cívicos* continued throughout the 1980s and succeeded in raising the related issues of responsiveness by the state and the devolution of powers to freely elected local governments.

This sense of a new politics, in which the people, and especially the excluded sectors, assume an active part in reconstructing the public domain, helping to create a political space suitable for hammering out the policies that will sustain an alternative development, is by no means an invention of the intellectuals I have quoted. For instance, the Political Commission of the First Extraordinary Congress of the Bolivian Federation of Agricultural Workers' Unions in July 1988 begins its report with the following declaration:

> We, the communards [i.e., members of the traditional peasant communities], are the majority of Bolivia's population (*los comunarios somos la mayoría de la población Boliviana*). Since the colonization of our continent (by the Spaniards), we have lived in hunger and misery. Exploited economically and oppressed culturally.
>
> National independence did not change our situation, which began to change only with the revolution of 1952. With *campesino* mobilization, we forced the MNR government to pass the agrarian reform legislation, and we gained our civic rights (*conquistamos nuestro derecho cívico*). (CSUTCB, 1989, p. 23)

"And we gained our civic rights." This is the new rhetoric of the popular sectors in their struggle for inclusion. Intellectuals merely register and interpret the new facts of political life in Latin America.

Admittedly, there is a utopian bias in the writings I have quoted. A greatly strengthened political community is asserting itself here against the deeply ingrained authoritarian and oligarchical tendencies in Latin American society (Veliz, 1980). People power may not be strong enough to overcome the limited democracy of the few who control the accumulation sectors of the economy, but the writers I have cited are not merely projecting another future for their countries; they are also witnesses of emancipatory struggles and everyday resistance.

The new energy of civil society among the excluded comes from having been oppressed for centuries, from having been silenced, from the daily struggle for survival, and from the pervasive feeling that the state has abandoned them, that they no longer count in the political project, and that they must take charge of their lives or become even more marginalized and oppressed.

The time when people looked hopefully to the state to resolve their problems has passed. They have learned that the state is neither all-powerful nor greatly concerned with their life situations. Economic policies imposed by the international banking system and now embraced by a growing number of conservative national politicians signal a return to growth – income maximizing and "trickle-down" policies that, even though they have rarely worked in the past, have begun to recapture the high ground of hegemonic rule. The result is quite the opposite of a "solution" to the problems of livelihood among the poor. Small peasants are made landless by these policies, and the popular sectors work "informally" even as real incomes continue to decline. And so the poor seek political empowerment to alter the balance of forces that have kept them excluded from a polity in which they now insist on their citizen rights.

Small Is Beautiful, But . . .

The typical example of an alternative development in practice continues to be a microproject.[1] It may be a shelter belt in the Sahel, the introduction of alternative technology into African food systems, traditional medicine in India, pig-waste recycling to produce methane gas in China, small-scale irrigation projects in Tanzania, credit programs for micro-enterprises in Bangladesh, and agricultural production cooperatives in Bolivia. Local-action projects of this sort typically bypass the state or else exist, barely noticed, on its margins. In any given country there may be dozens, hundreds, and in large countries even thousands of "alternative"

---

[1]   Such projects are described in Newell, 1975; Hirschman, 1984; Carr, 1985; Reid et al., 1988; Annis and Hakim, 1988; and Martens, 1989.

Table 7.1   Small-scale alternative development projects vs. typical large-scale
mainstream development projects

Alternative projects	Mainstream projects
Financial assistance goes directly to the poor	Financial assistance goes to the state
Relatively inexpensive, especially in terms of foreign-exchange requirements	Relatively expensive in terms of foreign-exchange requirements
People-intensive; fact-to-face interaction essential	Capital-intensive
Appropriate technology, often as extension of existing practices	Advanced technology, usually imported from abroad and displacing existing practices
Flexible management (changes possible in course of implementation)	Bureaucratic management (once committed to a course of action, changes are difficult to make)
Fine-tuned to local conditions	Procrustean: what doesn't fit must be "cut off"
Oriented toward mutual learning between external agents and local actors: transactive planning	Top-down technocratic planning; little learning occurs
Control for negative side effects relatively easy and quick	Control for negative side effects are delayed
Short start-up time	Long start-up time

projects.[2] In virtually every case these projects' aim is eminently practical:
they respond to a specific local need, their methods are experimental, and
their immediate results are often encouraging. But in a mainstream perspective, they are regarded as contributing little to economic growth and
capital accumulation; they do not count as "development" writ large.
Rather, they are seen as a form of poverty alleviation and, in some
ultimate sense, an inexpensive means of social control.

On the other hand, when we compare alternative projects of this sort
with the megalithic undertakings typical of the mainstream approach to
development – a major harbor development scheme, a giant power plant,
a large-scale irrigation project, an airport, a new aluminum smelter, a
metropolitan subway – the apparent advantages of small over large are
hard to contest. What is it that makes small beautiful?

In the schematic comparison of table 7.1, small projects would seem to
come out ahead. But there is another side to this story of "small and
large," for "small" undertakings also have serious drawbacks.

1.   Some "big" projects are essential to any kind of economic development and are not always divisible into smaller units. Small irrigation

---

[2]   For numerous Brazilian examples of innovative local action, see RECEM, 1986.

works may sometimes be preferable to large-scale projects, but hydro-electric power for industrialization requires big dams. They may be built in stages, but in most situations big dams are unavoidable. All that can reasonably be done is to minimize their negative impacts on the environment and on resident populations in the flooded areas who need to be resettled. Feasibility studies must take their social and environmental costs into account.

2.   The impacts of small alternative projects are primarily local: their significance is at the micro level. To reach all of the needy population, however, small projects would have to be replicated thousands of times. But replication on this scale is difficult and may well be impossible. The point is well argued by Devaki Jain, one of the founding members of the Indian Association of Women's Studies and currently director of the Institute of Social Studies Trust in New Delhi:

> There are several reasons why . . . successful micro-level projects are not generalizable. One is the *charisma* and *dedication* associated with the "first" experiment which usually cannot be replicated. Another is that the financial and ideological *investment* put into the original is often missing or hard to duplicate. A third is that certain cultures absorb what others cannot. My view is that the inability to replicate stems from all of these, and more. It is the innovative process itself that generates the first success which counts. The impetus, the consciousness raising, the leadership, the muscle and the "heavy weight" that developed the first project dissipates in succeeding ones that seek to duplicate it. (1989, p. 76)

Jain's answer is to strengthen local capacity for innovation. No two situations are ever alike, she says, and creative responses must grow out of an intimate understanding of resources, constraints, and political will.

3.   The transaction costs of many small projects are generally higher than for one big project. It is for this reason that both international banks and national ministries prefer "few and large" over "small and many."

4.   Alternative development projects are inherently difficult to coordinate. This is especially true when they are promoted, as is often the case, by multiple foreign donor agencies with different ideological and technical orientations. On the other hand, when they are obliged to work through national ministries and are thus subject to coordination, they may lose their innovative character. Loss of capacity for innovation may be the price exacted by conformity with bureaucratic norms.

There are no easy ways out of these dilemmas. Small works because it is small, autonomous, and tailor-made. Painted on a larger canvass, small may not work at all. And to this extent, it isn't an alternative development.

This conclusion corresponds to a project view of an alternative development. But increasingly networks, coalitions, federations, and confederations of popular organization and nongovernmental organizations exist that extend to the entire metropolis, to the nation, and even to multination regions. These ensembles serve as vehicles for information exchange, technical and political support, and political lobbies, establishing the civil society of the disempowered as a significant actor outside the traditional system of party politics.

Recognition of the need to expand the territorial scope of alternative projects – to "scale up," as it is called – and to devise organizational support for local project work was recently brought to the attention of donor communities by Sheldon Annis of the Overseas Development Council (Annis, 1988; Morgan, 1990). Writes Annis,

> In Latin America . . . a process [that maintains the virtues of smallness but at the same time reaches large numbers of people, transfers genuine political power to the poor, and provides high-quality social services that are delivered by permanent, adequately financed institutions] is already taking place. Every Latin American country is now interlaced with a thickening web of grassroots organizations.[3] These organizations are increasingly intertwined not only with each other but with the state. As a result, a policy built upon the idea of large-scale, small-scale development – something which might have appeared naive or whimsical just a few years ago – is emerging as a serious choice for Latin America in the 1990s. (1988, p. 210)

Annis argues that if an alternative development is to be sustained, it can no longer avoid the state. States make policies and command resources. They can do things that grass-roots movements cannot. The new scaled-up organizations to which Annis refers help to articulate civil society and state at all relevant levels of decision-making.

I will return to Annis's argument later; here it is sufficient to point out that while scaling up may be desirable from the standpoint of conventional effectiveness criteria in project management, it poses considerable dangers. Scaled-up organizations acquire bureaucratic features, power tends to drift upward, professionalization (which is almost always disempowering) takes over, and cooptation of popular organizations by a powerful state is likely.

There are, as always, exceptions to these dire predictions. Mexico's CONAMUP (National Coordinating Body of Mexico's Popular Urban Movement), with an estimated million member, has struggled mightily to retain its informal internal structure, its reliance on moral incentives, and

---

[3] According to the Interamerican Foundation (1990), there are now more than 11,000 nongovernment organizations (NGOs) in Latin America and the Caribbean.

its autonomy from the state. But CONAMUP is a "people's organiza-tion" that sees it role limited to social mobilization and public advocacy. In some ways, it behaves like a labor union, except that its focus is on housing rights and related issues rather than on wages and working conditions. It does not seek to preempt the state in what it perceives to be the state's business any more than labor unions seek to preempt manage-ment prerogatives. CONAMUP is first and foremost an advocate of people's rights (Moctezuma, 1990).[4]

Spontaneity Is Not Enough

Sheldon Annis refers to proliferating grass-roots organizations as "self-organizing systems" (1988, p. 210). He seems to imply some sort of "natural" or "spontaneous" process of social organization. But in fact the likelihood of a truly spontaneous organization of the poor is very small. The only unmediated action among disempowered households is mutual help and an occasional outburst of protest, such as Colombia's *paros cívicos*. But precisely because they lack formal organization, pro-test movements are also easily contained. Local leadership may be coopted, state responses to social demands may be predicated on the promise of community compliance, and more overtly repressive measures may be used to discipline both the community and its leadership. The various means available to the state for dealing with popular protest movements (usually defined as popular "unrest") are well understood and effective.

Moreover, spontaneous popular action generated from within village or barrio is rarely innovative but tends to select from a known repertoire of actions. There are many reasons for this, including poor people's need to minimize the risks to themselves, their shortage of "surplus time" for searching out and testing innovative responses to problems, their lack of innovative leadership, and inadequate financial, material, and technical means. If the practice is to be innovative – and continuous social

---

[4] For a similar example for India, see the story of SPARC (Society for the Promotion of Area Resource Centers), which works in housing advocacy, women's issues, and drug abuse. Like CONAMUP, SPARC works in a process-oriented mode. With its tiny budget it operates out of a tiny office with funding coming from the Indian government (including state and local) and international sources. But since its inception its budget has doubled every year. "The original budget [in 1984] was £4,900; now this has grown to between £130,000 and £148,000. Despite this, SPARC still operates like a household budget, with money being withdrawn weekly and paid to the workers to cover expenses. With the help of a computer, it is possible to continue working like this" (*Environment and Urbanization*, 1990, p. 97). One can only wonder how long informal arrangements can last, even with the help of a computer, should SPARC continue to grow as it has in the past.

innovation is a basic requirement of an alternative development – the rhetoric of spontaneity must be abandoned.

The typical alternative project is generated with the help of outsiders known in development-community jargon as external agents or in French as *animateurs*. External agents generate an organized response on the part of community groups to a new challenge. Their basic task is to "animate" – that is, to blow the breath of life into the soul of the community and move it to appropriate actions. They are meant to "spark" endogenous change "from within," not to carry out the change program; this is a responsibility of the organized community.[5]

But the practice of external agents is often otherwise, and what appears as a spontaneous creation of the community may, in fact, have quite different origins. This story of Lima's *comedores populares* (popular dining halls; similar to Chile's *ollas comunes*) is told by Susan C. Stokes:

> In one large section of Independencia, a popular district of Lima's "northern cone," there existed by mid-1986, 21 *comedores populares*, all of which were members of two separate dining hall "federations," one associated with a local training center, the other with the parish. In only two of 21 cases did local women independently start up their own dining halls; in general, the initiative was in some substantial measure that of nuns or the

[5] An outstanding example of "animation" in this sense comes from Bombay and is worth recounting here. By our definition SPARC is an external agent. Its work is informed by a well-articulated philosophy encompassing three points:

1   Those who face the problem are best equipped to identify the elements of a workable solution.
2   Processes not products create movements for change.
3   Informed participation is crucial for movements to be sustained and survive.

SPARC's operating methodology is derived from these general principles and can be summarized in six discrete steps:

1   Locate the central features of the crisis as identified by the community facing it.
2   Understand how the state perceives that crisis.
3   Share this insight with the community and debate the formulations of elements necessary for a solution.
4   Create an information base from participatory research.
5   Initiate professionals to take part in formulating alternatives with the communities.
6   Initiate a campaign for change: mass demonstrations, publication of information, and workshops; negotiate meetings with government.

"A central feature of SPARC's work is to create structures which reflect what people want. SPARC sees its functioning as a means towards achieving this, and all activities which it takes up must link directly back to this. The organization has attempted to develop internal mechanisms which reflect this philosophy and allow for its realization" (*Environment and Urbanization*, 1990, p. 95).

center's "promoters." And in all cases, the nuns and advisors were extremely active in providing an on-going orientation to the organization: they were not "crushingly authoritarian," but they were absolutely vital to the dining halls' survival. More important, these outsiders were centrally involved in keeping alive the communitarian ethic of the dining halls. They viewed the dining halls as models of how the poor could overcome the problems facing them collectively, and for the equitable distribution of food and other resources, models which were seen as transferable to the society as a whole. To the extent that this egalitarian ethic and form of organization is what analysts really mean when they speak of "autonomy," then it must be underscored that such popular-sector "autonomy" is due in large measure to the influence of non-popular-sector actors.

The view of independent *comedores populares* as self-financing contributes to the fiction of dining-hall autonomy. Virtually all of Lima's dining halls have two basic sources of income: one, in the form of foodstuffs, from CARITAS and other charitable organizations; the other, the cash income from "cuotas" or sales of meals to dining hall members.... The principal source of food donation is CARITAS which, according to one economist, provides a full sixty percent of the value of the typical dining hall's meals. (Stokes, 1988, pp. 6–8)

In Latin America as elsewhere, there is no shortage of "external agents." They include both domestic nongovernmental organizations (NGOs) and foreign private voluntary organizations (PVOs), as well as informal church groups, representatives of political parties, young professionals, university students, and even the state acting through some office of community development, the national health service, or similar organization. These groups form in response to a variety of impulses, and only some of them have "service to the poor" or "collective self-empowerment" as their leading motive. Religiously motivated groups may be interested in making converts, political parties in getting votes, and state agencies in securing control over civil society.

Franz Barrios Villegas, who works closely with Bolivian peasants, has developed a typology of NGOs that is relevant here. First, he says, there are those NGOs that completely identify with the development model and the mission of the existing government. Then there are those that describe themselves as apolitical but are socially progressive and jealously guard their critical independence vis-à-vis both the state and opposition parties. Finally, there are the NGOs openly in opposition to the government and politically aligned with the Left (Barrios Villegas, 1987). We may call group 1 parastatal, group 2 professional, and group 3 politically progressive. Writing from a Bolivian perspective, the author proposes to work toward what he calls a minimum consensus between groups 2 and 3: "[We must] arrive at a common base of strategic action that will give unity and coherence to our partial efforts. Though we maintain our separate identities, we must begin to synchronize our actions toward the single goal of a structural transformation [of Bolivian society]" (1987, p. 6).

One conclusion we may draw from Barrios Villegas's typology is that relatively few NGOs are both apolitical and committed to working with the disempowered at the micro level. In one form or another, most Latin American NGOs assume a political position on the ideological spectrum, either in defense of the existing power structure or against it.[6]

Nongovernmental Organizations in the Democratic Transition

Latin America's nongovernmental organizations may be considered a direct expression of civil society. At a meeting of 30 NGOs from nine Latin American countries in January 1987, this statement was adopted: "In terms of relations with other social actors, the relationship with the state has been accorded priority, based on the assumption that the division between civil society and the state should not be seen as absolute. The Centers define as one of their functions permanent denunciation and criticism. But the Centers also consider that they are in a better position than state entities for developing creative proposals that respond to the most fundamental social problems" (Landim, 1987, pp. 32–3).

---

[6] The nomenclature of NGOs has not yet settled into a consensus. Various typologies have been proposed (Korten, 1987; Garilao, 1987; Barrios Villegas, 1987). Here are some distinctions I have drawn:

1   Popular organizations are nonprofit, nonpolitical groups from within the civil society of the poor. For the most part, they are funded by membership dues. Examples include CONAMUP and Mexico City's Asamblea de Barrios.
2   NGOs are officially registered professional groups whose university-educated core staffs focus their work on communities of the disempowered. NGOs are funded through private dona- tions, Private Voluntary Organizations (PVOs) and the state. Following Barrios Villegas, NGOs may be further divided into parastatal, professionally oriented, and politically progres- sive organizations. They may engage in a variety of tasks, from project work to process- oriented work, action research, and pure research. Examples include Lima's CENCA (action research) and DECSO (pure research), Santiago's CIMPA (environmental research) and PET (action research on employment and work), and Bogotá's CINEP (action research and training) and Construyamos (project work).
3   PVOs are foreign NGOs. Some of them have in-country operations; others are primarily donor organizations; a few do both kinds of work. They are typically global in scope and operate with very large budgets, and they may also do contract work for international aid agencies. Examples are CARE, Oxfam, the Friends' Service Committee, the World Council of Churches, the World Wildlife Fund, and Conservation Foundation.
4   Nonprofit, socially oriented business organizations, such as India's state-of-the-art Develop- ment Alternatives (New Delhi), which designs, manufactures, and sells village technologies.

In this declaration of principles, NGOs appear as organizations ("centers") clearly outside the state's domain. More precisely, they stand *in opposition to* the state. But accepting existing realities, they are also prepared to *work with* the state in developing creative proposals that will benefit the disempowered, for "the division between civil society and the state should not be seen as absolute."

Although Latin America's NGOs may not be overtly political, they are politicized and jealously guard their individual autonomy. According to Landim, "Independence [is] one of the cornerstones of their discourse; and they are oriented toward individualization, as nuclei of power in civil society" (1987, p. 33). He concludes "Between highly committed volunteer service and professionalization; between the Church, universities, and political parties; between Christianity and Marxism; between a conspiratorial brotherhood and institutional relations; it is amidst these contradictions that the agents of these *sui generis* entities make their way" (1987, p. 33).

Like other organizations of civil society, NGOs reflect the political culture of their respective countries. Compared with their Asian counterparts, Latin American NGOs tend to be more political. Ernesto D. Garilao, for example, speaking for the Philippine Business for Social Progress, refers to NGOs as an economic sector:

> "Sector" is not used loosely here. It is used precisely to denote that NGOs as a group have a distinct socio-economic-political function. As such, they can be juxtaposed with the traditional sectors of the economy, the public and the private sectors.... As NGOs expand and professionalize their services, and attempt to bring in more of their population from the margins of society, they are in fact creating a new service industry – the social development industry. (1987, pp. 115–16)

Garilao's description may be accurate for the Philippines, but it is not an account of how things are done in Latin America. Of course, here too "intermediate" NGOs are beginning to appear (Sandoval, 1988; Albertini, 1989). But so far at least the relation between state and civil society has been one of constrained mistrust if not outright hostility, and where the two have worked together, it has generally been a case of antagonistic more than friendly cooperation (Sanyal, forthcoming).

Let us look more closely at the actual institutional arrangements. Figure 7.1 presents two models of NGO-state relations based on observations in Chile. Both focuss on the local community.[7] Model A represents

[7] It should hardly be necessary to note that my reference to community in this context is not intended to suggest social homogeneity, consensus, and harmony, characteristics that frequently get confused with local-level work. The poor may be largely disempowered – and it is their

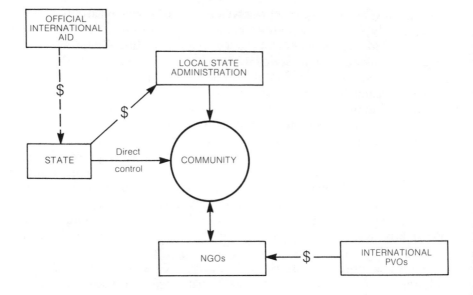

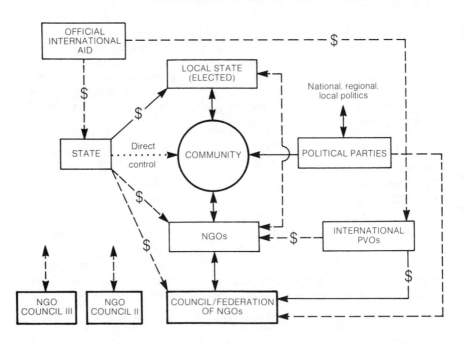

Figure 7.1 Two models of state-NGO relations

the relations between state and NGOs during the Pinochet dictatorship (1973–90). Model B conforms to the relations that can be expected during the coming years of the democratic transition. Outside Chile, similar features can already be observed in Peru, Colombia, Ecuador, and Bolivia.

Model A is by far the simpler one. The central state controls the local community in two ways: through direct action, via its own community development efforts and police action, and through the administrative arm of the local state.[8] During the period of the military junta, NGOs – supported by West European PVOs, which in turn received their funding from liberal church groups and center-left political parties – fostered a spirit of democratic resistance to the regime, even as they worked with so-called *comunidades de base* (base communities) to resolve urgent problems of both material and spiritual survival (Campero, 1987; Lehmann, 1990). As NGOs promoted cooperative solutions, self-help, and participatory democracy at barrio and village levels, the national state was kept at arm's length. It was the "enemy."

Paradoxically, the localist philosophy of NGOs found sympathetic echo in the neoliberal doctrine of the military regime. The "New Right," it turned out, was not necessarily opposed to community-based action among the poor, as long as it remained encapsulated within the local sphere and did not aspire to political power and claim-making as well. In neoconservative doctrine, the so-called minimalist state (minimalist in the sense of not interfering with the "free" market) goes hand in hand with a strong, organized civil society at the local level (Nisbet, 1953; Nisbet and Perrin, 1977; Berger and Neuhaus, 1977; Novak, 1980). Democracy interpreted as the marketplace of politics – that is, as a consumer democracy, relying on political representation – is viewed as homologous with "free enterprise." The model and practices of a consumer democracy resolve the potential contradictions between popular democracy and capitalism. Moreover, poverty-alleviation programs save the government a good deal of trouble and ease its efforts to locally manage popular discontent.

As for the organized "community," it is always glad to collaborate

---

disempowerment they have in common – but looking closely we see differences even in the form and degree of their disempowerment. More than that, civil society – households forming different "communities" – are divided among themselves by kinship, social class, ethnicity, religion, gender, rural versus urban base, occupation, and political tradition, all of which renders any idea of "harmony" at best irrelevant and at worst harmful. People work together for whatever reasons they have. The also fight amongst themselves, often bitterly and with no holds barred. "Community" is therefore used here for no purpose other than to designate the *local* sphere of action.

[8]   Pinochet abolished local self-government in 1973. Local communities were governed through the Ministry of the Interior; mayors were appointed by the president: There were no elected councils. All these provisions are now being changed to conform with democratic procedures.

with NGOs as long as there are practical benefits to be had. In the absence of a caring government in Model A, NGOs are the community's principal lifeline to the world beyond village and barrio. Typically suspicious of politics, community organizations are content to receive assistance from whoever is willing to provide it, *including* the state. According to Eugenio Tironi, the interests of *pobladores* (local residents among the popular sectors) are strongly particularistic and material (1987).[9]

The democratic transition Model B is considerably more complex. Local governments are now elected, and a multitude of political parties are contending for the "soul" (i.e., the votes and political support) of the community. Official international aid, which had been reduced during the latter phases of the military regime, is beginning to flow again, while PVO contributions – now that the political imperative is over – are gradually being cut back. Meanwhile, indigenous NGOs have learned to work together, forming coordinating councils or federations. Figure 7.1 shows three such federations, each corresponding to one of Barrios Villegas's categories: parastatal, professional, and politically progressive. But contrary to Barrios Villegas's hopes, the first two are gaining relative to the third, as the state itself now spends considerable sums of money on NGO projects. Coincidentally, official international aid also has begun to finance parastatal and professional NGOs (Korten, 1990).

As more and more professionally oriented NGOs are brought into the framework of policy-making by the state, the remaining NGOs, many of them aligned with opposition parties, are rendered highly vulnerable. Starved for funds, they may cease operations altogether. Those that survive will tend to concentrate on marginal projects, political education (consciousness-raising), and advocacy.

In the democratic transition model, disempowered people now have access to a multiplicity of political channels. Chile's new coalition government allocates between 4 and 6 percent of its budget to special poverty-alleviation programs. The antistate rhetoric of Model A is thus displaced by a more accommodationist discourse. And, in the face of political fragmentation, the state remains firmly in control.

Despite intense local activity, very little progress is made in terms of people's self-development. Unless economic growth *of the right* kind is

[9] During the early years of military rule that spread cancer like throughout South America from the mid 1960s onward, left-wing analysts had it that the new barrio-based social movements would become the historical carriers of a new revolutionary wave for a socialist alternative. Such was not the case, however, and the evidence pointed increasingly in another direction. What community people seemed to want more than anything else was what every Chilean also wanted – a job, a decent income, freedom from police harassment, and a little house with a garden. They were not interested in "social experiments" on a grand scale. They were the most reluctant of revolutionaries. All they wanted was justice. All they really asked for was to be "included."

revived and sustained at the national level, the poor will have few possibilities for springing the poverty trap in which they are caught. They will remain disempowered.

I don't want to give the impression that with the return of democracy in Chile, or anywhere else for that matter, the promise of an alternative development is lost for good. Far from it. The institutionalization of a democratic politics, especially if it reaches down to the local sphere to become a genuine grass-roots politics, represents a real gain. The return of elected local governments and the promise of barrio self-government through neighborhood councils with formal attributions would be a significant step forward (SUR, 1989). The minimum conditions will have been created to carry forward a politics that seeks imaginative structural reforms in legislation and policy with a view to breaking down the wall that keeps the disempowered safely out of sight in ghettos on the far periphery of the city.

A Different Politics

The models of state-NGO relations have taken us some distance into the politics of an alternative development. They do not, however, exhaust the varieties of social and political movements that exist throughout the southern continent. Here I would like briefly to discuss three examples of attempts at an alternative development on a scale larger than the locality.

The first concerns an unprecedented election victory that in 1984 carried Peru's United Left to power in the municipality of Lima. In the three short years it remained in office, the United Left attempted a radically different approach to the housing question, initiating five pilot projects. Central to the traditional way of city-building in Lima are *barriadas* (self-built housing settlements) on unoccupied land on the outskirts of the city. Land seizure and occupancy is a highly ritualized "ballet" of confrontational politics, usually ending in a peaceful process of urban land consolidation. Over the years, individual housing units are built, brick by brick, replacing the soft, temporary materials – cardboard, straw mats, tin roofs – of the original huts. Through persistent efforts, residents eventually gain title to the land. Roads are paved. Water lines are laid. Electricity is provided. Community facilities are built. The entire process may take 30 or more years – an entire generation. But finally, what had begun as a *barriada* will have connected with and become an integral part of the urban "web," undistinguishable from other low-income, already-consolidated areas (Hardoy and Satterthwaite, 1989).

In 1984 the Young Turks of the United Left were eager to change this pattern and to create human settlements more reflective of the group's

ideology. They wished not only to accelerate the process of land consolidation – compressing it into only a few years – but also to give it a
different physical appearance involving urban designs that, they hoped,
would promote neighborhood solidarity and remove the social stigma of
living in a *barriada*. The architects and planners of the United Left also
wanted to experiment with the *coproduction* of new settlements, involving a joint venture between state and local community – a public-private
partnership at the "popular" level.

Before attempting to assess the results of this experiment I should say
that three years is scarcely enough time to test the feasibility of an
alternative approach to low-income housing. At the end of 1986, when a
new city administration with a different political orientation took over
Lima, the United Left pilot projects were abandoned. Three years was
not enough time for effective social learning to occur. The different time
perspectives – a government looking for quick results and eager to prove
the superiority of its ideas, and a community accustomed to work within
the time frame of a human generation – exacerbated conflicts that, given
the situation, were probably unavoidable.

But let us listen to the conclusion of a detailed, careful evaluation of
two of the five pilot projects:

> It is undeniable that in Huayacán and Laderas del Chillón, the municipal
> projects [of the United Left] have produced . . . well-organized settlements
> through a technical assistance process of some consequence. They suc-
> ceeded in accelerating the successive stages by which a *barriada* becomes
> consolidated with the city, installing basic services, granting property titles,
> engaging in comprehensive planning, making novel proposals for citizen
> participation, etc. But on the whole, they could not overcome two critical
> conditions: to build a genuine alternative to the *barriada* and to counteract
> the *barriada*'s cultural image in the minds of its residents. . . .
>   . . . [T]he consolidation of the projects collided with the pragmatic cul-
> ture of the local residents who understood the *barriada* not simply as the
> seizure and permanent occupation of a site but as a process that, given
> certain conditions, would culminate in at least a kind of housing that
> would help to resolve a number of economic problems for the household.
> To be in possession of a house was to have a secure place for effecting
> changes [in the conditions of the household], a life space capable of
> expansion, and individually provided services – conceptions that, on the
> whole, ran counter to the technical proposals [of the United Left] which
> stressed collective answers to individual problems. (Calderón and Cárde-
> nas, 1989, p. 127; my translation).

The attempted coproduction of housing (involving innovative mechanisms of participatory planning and building) foundered on these differing

conceptions of what, respectively, people and government each wanted. Given enough time, these differences could perhaps have been worked out. But the necessary time did not exist.

The general lesson we may draw from this example is that any government that expects the organized community to be a part of devising a comprehensive housing solution must be willing to listen to and learn from its residents, whose capacity for passive resistance is great, and to follow their lead. But this requirement also implies a need to rethink professional planning roles that, in the Peruvian case, apparently had not occurred.

A very different example comes from Mexico City. The earthquake of 1985, which left large sectors of the old central city in ruins, gave rise to the most articulate, organized citizen movement since the revolution. Massive protests obliged the government – which had other intentions – to reconstruct the demolished areas. According to political scientist Susan Eckstein, "About 70 percent of the people have relocated in new housing on their old land sites. A sizeable fraction of those who moved, did so by choice. Furthermore, like the *vecindades* they replace, the new units are built around inner courtyards and are no higher than three stories. For many center city people, the new housing is an improvement" (1988, p. 231). She then goes on to say, "The new housing that has so completely transformed *El Centro* physically is a byproduct of 'popular' protest and other defiant acts. The government most probably would not have built housing for residents, and housing that was at least somewhat responsive to local lifestyles, had the people been quiescent. Indeed, the evidence suggests that had the government had its way, the area would have been bulldozed, and the land made available for more profitable commercial use" (1988, p. 263).

The full story of this protest movement has been told by Professor Eckstein, and I don't wish to repeat it here. Instead, I would like to focus on an offshoot of the original movement that, beginning in 1987, continued the struggle for housing and extended it beyond the central-city areas that were damaged or destroyed by the earthquake.[10] Calling itself La Asamblea de Barrios de la Ciudad de México (the Mexico City Assembly of Neighborhoods), this offshoot movement quickly evolved into a federation of local housing cooperatives and other groups throughout the metropolitan area, representing some 50,000 households – about 250,000 people – who were *sin techo* ("without a roof").

From the beginning, the Asamblea was embarked on a dual struggle: for the right to decent housing and for the right to "a city for all" (*una ciudad para todos*). The latter was a claim for democratizing urban

[10]   The following account is based on Rivas, 1989.

governance at the level of the Federal District, an area that continues to be administered by the central government without citizen participation or elections.

Building on widespread disaffection from the ruling party – the Institutional Revolutionary party (or PRI by its Spanish acronym), which has managed the country's affairs largely unchallenged since the 1930s – and adopting unorthodox methods of publicizing its struggles, the Asamblea became an effective advocate for 43 of Mexico City's organized housing groups: first-time homeseekers, long-term applicants for housing loans from government agencies, barrio organizations, residents of public housing estates seeking specific improvements in services and repairs, tenant unions, youth groups, women's groups, and street vendors. The Asamblea depended largely on informal processes and networking for its activities and has so far avoided building a bureaucratic infrastructure. Weekly meetings kept up popular enthusiasm and commitment to the drawn-out struggle; street demonstrations and marches were organized; cultural festivities added to the spirit of conviviality ("protest can be fun!"). Effective use was made of street theater, and a genuine creation of popular culture materialized in the persona of the flamboyant Super-Barrio, who, decked out in cape and tights like the American Superman, assumed the role of popular defender of the poor.[11]

Its founding, months before the presidential election in 1988, coincided with the formation of a breakaway faction of the PRI, headed by Cuauhtémoc Cárdenas and Porfírio Muñóz Ledo. Calling itself the Democratic Revolutionary Party (or PRD by its Spanish acronym), it declared its opposition to the PRI, challenging its political hegemony. As a party of the Left, it was only natural that it should champion a democratization of governing structures that, in more than half a century of single-party rule, had become deeply encrusted with graft and nepotism and would seek to hold on to power by any means, fair or foul. As it turned out, Cárdenas, though claiming fraud, lost the presidential election. He continues, however, to be a major political force in the opposition.

From the outset the Asamblea de Barrios was to forge strong linkages with Cárdenas's and Ledo's PRD. Its demands for "a city for all" and a radical democratization of urban governance largely coincided with PRD's own platform, while its massive, colorful mobilizations could effectively be used by the opposition to embarrass the government. On the other hand, PRD, even in opposition, would lend a political voice to

---

[11]   Whereas America's Superman comes to the rescue of imperilled individuals, Mexico's Super-Barrio comes to the rescue of the poor collectively.

what otherwise would have remained a simple case of housing advocacy, easily coopted by the government.

The other side of this coin, of course, is that the PRI government is not likely to be financially generous toward a housing advocacy movement that has formed a working alliance with its most serious political opponent. Today the Asamblea is walking a tightrope between political objectives and bread-and-butter advocacy of housing claims. If it only pursues the first, it will probably lose most of its support in a short time. As an abstract idea, democracy is not of great interest to the disempowered. Their practical concern is housing policy. But without its assertion of citizen rights in the concrete sense of "a city for all," its dramatization of the city's deteriorating housing situation would lose resonance. Maintaining a dynamic equilibrium between the two directions of its struggle is a difficult and challenging task.

My final example is the Bogotá, Colombia, housing NGO, Construyamos, or Let's Build![12] Its origins are revealing. The first self-help housing schemes in Colombia were established more than 20 years ago, and by 1982 more than 500 independent housing cooperatives were in existence. On their own, however, these mostly small cooperatives lacked adequate technical, professional, legal, and financial resources. With the support of the Colombian government, therefore, the *autoconstrucción* (self-built housing) movement was encouraged to form a national association that would facilitate the exchange of experiences and resources and to represent community-based organizations before government agencies, major financial institutions, and international aid agencies.

At a government-sponsored congress in July 1982, grass-roots activists gave voice to a litany of problems that beset most housing projects at the time. What was needed, they argued, was better representation and coordination at the national level. They also thought that an effort should be made to integrate the hundreds of organizations – housing cooperatives and so on – dispersed across the country. Early the following year Construyamos was formed in response to these demands, with a mandate to promote self-help housing, to lobby for much-needed reforms in housing legislation, and to provide adequate support services to local groups.

On the invitation of the housing advisor to the president of the republic, an international mission, composed of members of the Royal Architectural Institute of Canada (RAIC) and funded by Canada's official aid agency (CIDA), attended the founding conference of Construyamos in April 1983. Based on its findings, the mission recommended the

---

[12] The following is based on preliminary field research by Maurício Salguero Navarro.

development of a three-year program of institutional cooperation between RAIC and Construyamos, with financial support from CIDA. Under the terms of an agreement, Construyamos would receive core funding for its national office, committees, and field operations, while RAIC would provide the technical and management advice Construyamos required to develop its organizational network and field operations. In subsequent years, the project was renewed twice, with final funding scheduled to run till June 1992.

What does Construyamos do? How does it define its mission? Six specific objectives outline its agenda:

- To stimulate the participation of project beneficiaries in identifying their shelter requirements, formulating appropriate alternatives, planning, executing, and maintaining their housing projects, and evaluating the results of their efforts.
- To facilitate the access of project beneficiaries to related services, such as education, training, employment, recreation, and health.
- To promote the full participation of women within the self-help housing process and to ensure that women assume leadership and decision-making roles within community-based housing organizations as well as within the internal organization of Construyamos.
- To effect significant improvements in government legislation and bureaucratic procedures relating to community development and the production of low-income housing.
- To improve the efficiency of the self-help housing process in order to reduce the time and costs involved in delivering *autoconstrucción* projects and ensure the lowest income groups access to adequate housing, infrastructure, and services.
- To consolidate the internal organization of Construyamos at the national and regional levels and to foster improved coordination with Colombian NGOs involved in housing and community development.

What appears clearly from this list of objectives is that Construyamos is much more in tune with the requirements of institutional aid funding than the more loosely organized and politicized Asamblea de Barrios in Mexico City, whose minimal funding comes largely from its own membership. Whereas the Asamblea is thus accountable to its collective membership, Construyamos, at least in a financial sense, is strictly accountable to the Canadian government via RAIC and CIDA. As a consequence of its institutional role (and true to its origins as an agency formed with government blessing under Colombian law), Construyamos is far more bureaucratically organized than the loose network structure that seems to work well for the Asamblea. Unlike the Asamblea, the Colombian NGO is nonpolitical, serving partly as a lobby at the national level while offering a range of technical services to its membership organizations. (Construyamos housing projects are financed by regular Colombian credit institutions.) Because of these characteristics, Construyamos is very likely to become a fixture in Colombia's low-income housing market.[13]

Construyamos is a notable instance of what some are beginning to calling third-generation NGOs. Korten (1986, 1987, 1990) recalls the origins of nongovernmental organizations in the relief and welfare field. Then, in the 1970s, came the now familiar agencies working the vineyards of small-scale, self-reliant, local development. More recently, a third-generation "strategy" has emerged. Its objective, according to Korten, is "sustainable systems development," and its focus is the region or nation. Involving the "strategic management" of "all public and private institutions that define the relevant systems," its primary agent is a new kind of intermediate organization. Korten correctly perceives that third-generation strategies will mean less direct involvement of NGOs at the local level. Instead, he envisions a new kind of NGO that will assume a "foundation-like role ... directing ... attention to facilitating development by other organizations, both public and private, of the capacities, linkages, and commitments required to address designated needs on a sustained basis" (Korten, 1987, p. 149).

By the end of the decade, Korten followed through on his idea, founding the People-Centered Development Forum for the purpose of "encouraging and strengthening voluntary activities toward the realization of a people-centered development vision." The leaflet from which this quote is taken describes the body as "a values driven voluntary network of activist intellectuals who are helping to shape the directions of an emerging people's development movement." It was perhaps a logical next step in the evolution of development NGOs, which in a neoliberal climate of government retrenchment have become something of a growth industry. Whether "activist intellectuals" led by a California-chartered nonprofit corporation can also become the catalyst of a genuine "people's movement" by seeking to "strengthen the movement's clarity, direction, identity, and legitimacy" remains an open question.

---

[13] Some might wish to argue that Construyamos is not an NGO at all but what we have here called a popular organization. If not for outside funding, however, the body would not exist at all, and its formal and increasingly bureaucratic character places it closer to an NGO profile than to a more loosely hung popular organization. Still, such hybrid forms as this may stand the best chance for survival.

Implementing an Alternative Development: The Lessons

This chapter, which addresses the broad question of implementing an alternative development, has shown that, when we enter the terrain of *practice*, the conceptual purity of an alternative development must be abandoned. Many practical roads can lead to an alternative development. Most of them are not direct; they meander like paths through the thicket of politics. We may now summarize and in some cases expand on our findings in the form of ten conclusions.

1. *To be small and local is not enough.*   So-called alternative projects are prospering at community levels. There are countless examples of useful and participatory actions that directly improve the conditions of life and livelihood of the poor. But relative to the massive needs of the poor throughout the world, they are a mere drop in the bucket. Primarily, they are conceived by their sponsors as poverty-alleviation projects. Political and economic empowerment – the inclusion of the excluded – is not generally part of a people-centered development approach.

2. *The state remains a major player in an alternative development.* Advocates of "small and local" have tended to see the state as an adversary. The state reciprocated the honor, largely by ignoring development efforts sponsored by PVOs and NGOs in both urban and rural localities. Typically, the state's attitude was one of benign indifference. It tended to regard NGO efforts as a relatively cheap and convenient form of social control. But the moment it understood that "small and local" was not sufficient to contain popular "unrest" – the label with which the state usually designates protest movements – it reappeared as a major actor. A number of NGOs have begun to realize that they must work *with* the state rather than against it, and that the state may be influenced through lobbying, the electoral process (hence the need for an inclusive democracy), and, in the last analysis, radical reforms that will render the state more responsive to the claims of the disempowered.

3. *Spontaneous community action is limited in scope. External agents are needed as catalysts for change to channel ideas and resources to the community and to serve as intermediaries to the outside world.* The external agency of popular organizations and NGOs is clearly needed to initiate community efforts at local self-empowerment. But an alternative development cannot remain encapsulated within impoverished communities. *To be local is not enough.* If the majority is to benefit, higher levels of development action are needed. And that is why both popular organizations and NGOs have begun to form networks and coalitions in an attempt to reconcile the particularity of *lo local* with the general problem of regional and national development. In many ways the

territorially bounded region suggests itself as the ideal-sized unit for articulating local with national/transnational interests. Proponents of an alternative development should increasingly seek to direct their efforts at the regional level, linking into the great tradition of cultural regionalism that has shown itself to be effective in mobilizing popular energies around issues of self-development, resource conservation, and greater political autonomy (Mumford, 1938; Weaver, 1984; Friedmann and Forest, 1988; Stöhr, 1990).[14]

4. *NGOs are beginning to "scale up" their operations and are coming to play intermediary roles between state and civil society.* In this process NGOs will inevitably change their character. They will become institutionalized; the numbers of central staff will increase; their interactions with state officials will become more frequent and formal; they will begin to receive subsidies or grants from the state for reallocation among local projects; they will spend more time in preparing documents to obtain resources, to evaluate their own activities, and to inform the state (and overseas PVOs) of what they do. One might call this process the "formalization" of NGOs. In exchange for material and political power, they will lose some of their independence and sacrifice direct, continuous contact with people's organizations. These scaled-up NGOs must now satisfy several masters: field staffs (often separately accountable to different PVOs), popular organizations that continue to be their main clients, major foreign donor agencies, and the state. They will find it difficult to be equally pleasing to all masters. In this process NGOs with an explicit political agenda, especially when opposed to the government, will tend to become marginalized.

5. *Because intermediary roles make NGOs less reliable as an effective advocate of the claims of an alternative development, the disempowered poor need to acquire a political voice of their own.* In the specific example of Mexico City's Asamblea de Barrios, we pointed out the precariousness of walking the tightrope extended between political and instrumental objectives. The Asamblea's links with a major opposition party, the PRD, created difficulties for itself in obtaining government support for its housing projects. Social movements such as the Asamblea,

---

[14] The question of scale has to be carefully considered, for regions vary enormously in size, depending on the country and delimiting criteria. A federated state in India may have tens of millions of inhabitants, while a Guatemalan province may number less than a hundred thousand. In the Indian case, the federal state territory may need to be further regionalized into "development blocks," while in Gualemala a regrouping of provinces into larger regional ensembles may need to be considered. Ideal-sized regions do not exist, and even small countries have regions capable of some degree of self-governance. In drawing regional boundaries, it is important to bear in mind historical patterns, existing institutional arrangements, cultural identity, and the territories' natural base in natural resources.

with their informal network character and imaginative use of opposition politics (street theater, cultural festivals, etc.), are nevertheless essential to raise an alternative development from its entrapment in small, local projects to a citywide and even national level of importance.

6. *A politically progressive state will do poorly if it undertakes direct-action projects of its own, replacing NGOs.* The example cited was that of Lima's United Left. Contrary to expectations, this group's relations with community groups in the pilot project areas turned out to be highly conflictive. This potential for conflict is inherent in the situation and suggests that even a politically progressive state may do better to leave project operations to the organized community working hand in hand with NGOs.

What a politically progressive state can do well, in principle, is (a) to create a political space in which the claims of organized society can be heard and appropriate solutions worked out; (b) to mobilize financial and material resources in support of alternative development programs; (c) to remove legal and other obstacles to the self-development of the poor; (d) to propose new legislation that will facilitate the process of self-development; (e) to launch new programs in support of community-based efforts; and (f) to provide an administrative framework supportive of alternative development approaches.

7. The approach to alternative development described in the preceding point *builds on people's own initiatives, with the state playing essentially an enabling, facilitating, and supportive role.* State agencies should prepare themselves accordingly to create the capability of responding to local initiatives rather than impose dramatic initiatives of their own.

8. *A social learning approach to an alternative development centered in the community has the greatest prospects for successful implementation.*[15] Even though the state is a major player in an alternative development, the *initiative* for projects should nevertheless be centered in the community. Direct community involvement is a delicate and time-consuming process. Experience has conclusively shown, however, that without community involvement projects are difficult to implement successfully and are even more difficult to maintain once direct project aid has ceased (Martens, 1989). Centering projects in localities (and regions) requires mutual learning, patient listening, and a tolerance for contrary views. It also requires one to think of the project as involving a process of social learning, with frequent assessments of what has been accomplished and what has gone wrong, and a willingness to make appropriate adjustments in the course of the implementation process itself. In all of this, it is the community that plays the central part: the "external agent" serves

---

[15] For social learning approaches in planning, see Friedmann, 1987, chapter 6.

as catalyst and mediator, and the state performs a series of essentially complementary, supportive roles, including that of establishing the basic rules of the game.

9. *The popular sectors of civil society need jealously to protect and expand their autonomy vis-à-vis both the state and NGOs.* The state's basic attitude toward the popular sectors is to remain watchfully attentive. In a democratic transition, direct controls (police action, administrative measures, intimidation) are in part replaced by the softer paths of indirect control, such as poverty-alleviation programs and a greater willingness to negotiate solutions. But unless the community has the means available forcefully and convincingly to assert its claims, its struggles for inclusion will amount to very little.

Autonomy can be furthered in several ways. Territorially defined communities may be formally represented by *juntas de vecinos* (neighborhood councils) with authority over a wide range of issues such as human rights, local security, housing and urban infrastructure, quality of the environment, the needs of the poorest, the quality of local education, the quality of community health services, local production for use and for the market, popular culture, and natural disasters. For all these matters, the local community must have both a forum for political debate and an authoritative voice. The objective would be to establish a meaningful and effective form of democratic territorial governance at the level of barrio and rural district.

Another way is the need for social advocacy movements, such as Mexico City's Asamblea de Barrios, which are beyond the reach of the state and make the poor both more visible and audible at higher territorial levels of governance. A third way is to strengthen the internal social networks of barrio residents and small peasants by encouraging the formation of interest-group organizations, including regional migrants associations, youth clubs, sports clubs, fraternal organizations, women's organizations, local improvement associations, tenant associations, ecology groups, cooperatives, ecclesiastical base communities, health organizations, school committees, and groups promoting urban agriculture, in the infinite variety of ways that civil society has for giving structure, coherence, and meaning to everyday life. A dense network of civic organizations strengthens community. It channels new information and resources to the community. It increases the community's stock of knowledge and makes it more adept in using it. It augments its political voice and increases its capacity to quickly organize around issues when they become critical. Building civic organizations is an empowering social process.

10. *The unity of civil society is to be found in its diversity.* Civil society gets stronger to the extent that it organizes around its own

concerns. Over the past 150 years, much passionate rhetoric has been wasted over the importance of preserving the "unity of the working class" and of strengthening its solidarity relations. It was an old-fashioned Marxism that addressed the "working class" in this way, as though its common material interests were also its most powerful incentive for an emancipatory political practice.

Throughout this study, I have purposely avoided speaking of the "working class" and have referred to the disempowered as the "popular sectors" and "small peasants." This language, I hoped, would preserve the real-life diversity of "one-half the population." The divisions of civil society by kinship, gender, ethnicity, religion, and so forth likewise point to powerful motivators for action. What diverse groups have in common is their poverty and relative disempowerment, but even poverty is shaded differentially. Landless peasants and small manufacturers in the informal workings of the urban economy have, in fact, little in common, although both are to some degree poor. If empowerment is the general answer to their common poverty, the means for accomplishing this must be adapted to the specifics of each situation. The *political* challenge, on the other hand, is to discover a formula capable of accommodating a variety of interests and claims within the context of an alternative development without simultaneously alienating important sectors of the middle classes, whose continued support for an alternative development is crucial. The poor will never become empowered if their gains come at the expense of these sectors, whose own location on the scale of (dis) empowerment is never fixed precisely and is always at risk.

Conclusion

In the course of my argument, I have used mainstream doctrine as a foil in elaborating the theory and practice of an alternative development. As we conclude this intellectual journey, it may be useful to take another look at both the mainstream and the alternative to it in order to be very clear about the issues at stake. We will also want to look at how the relationship between the two may be articulated in practice.

A recent essay by Tyler Biggs, Merilee S. Grindle, and Donald S. Snodgrass – all of them with the Harvard Institute for International Development – will help us to concentrate on the currently popular interpretation of mainstream doctrine (Biggs et al., 1988). The essay was presented at a seminar convened by the U.S. Agency for International Development in 1987 under the umbrella title "Including the Excluded in Developing Countries." Ostensibly addressing the question of what to do about the notoriously elusive "informal sector," Biggs and colleagues spell out the mainstream doctrine of the late 1980s with admirable clarity. The issue, they say, is market-efficient economic growth. More is better. You can't solve the problems of poverty without economic growth. The authors admit that there may be problems of maldistribution. But, they say, "the risk is here that governments may sacrifice too much economic growth for greater equity" (p. 160). The solution is "equitable growth," they say, but efficiency requires that successful enterprise be rewarded. "As agents of economic development, very small enterprises are, to put it bluntly, of little interest" (p. 170). Although they should not be suppressed "unless they pose a clear threat to public safety or morality," preference in policy allocations should be given to efficient medium-scale producers. Overall, there is too much meddling of government in markets. Policies promising big payoffs in the long term should be pushed through resistant political structures, including *exchange rate reforms* (devaluation), *reforms in trade policy* (eliminating and/or lowering tariffs to expose domestic enterprise to the chilling effects of international competition), *financial and fiscal reforms* (interest rates, credit controls, level and structure of taxation, all of them to favor capital accumulation), and *investment incentives* (rewarding high achievers in the "dynamic middle") (1988, pp. 167–9).

By this formulation, mainstream doctrine urges the strict application of an accumulation model of the economy. Industrialization is regarded as the key to structural change. As for the poorest, the basic recommendation is to leave them alone (unless their actions are considered "harmful" – one thinks of coca growers in Bolivia and Peru who have become a U.S. military target), and otherwise to rely on the familiar "trickle-down" process that theory predicts when labor markets are tight. According to the authors, "The cure for [economic] dualism [where the many earn little and the few earn a lot] is development [i.e., economic growth]. As the empirical economists have shown, dualism disappears as economies grow and generate sufficient demand for unskilled labor" (pp. 169–70).[16]

The contrast with an alternative development model could not be more striking. But it also poses a challenge to this model.

1   If mainstream policies aimed at maximizing economic growth within an international division of labor generate massive poverty, as I have argued, isn't an alternative development, as long as mainstream

---

[16] The faith in the curative powers of economic growth reached an absured rhetorical zenith at the 1990 Economic Summit in Dallas, Texas, when the national leaders of the seven richest and most powerful countries in the world came up with this gem in their final declaration: "We recognized that strong, growing market-oriented economies provide the best means for successful environmental protection" (*Los Angeles Times*, July 12, 1990, p. A-11).

doctrine prevails, merely "treading water" to keep the poor from drowning?

2   Examples of alternative development practice involve actions primarily at the local level. But can local and even regional actions ever "add up" to a satisfactory national development?

3   How are alternative development approaches related to the prevailing mainstream doctrine?

If an alternative development is to mean anything other than emergency actions to spread a safety net under the poor, these questions must be very seriously considered.[17] I would like to concluded this essay by attempting a reply.

An alternative development does indeed address the condition of the poor directly. It argues for their involvement in actions that will lead to their own empowerment. In so doing, alternative approaches acknowledge the existence of the poor and their rightful claims as human beings and as citizens. But the alternative is not limited to local actions warding off immediate threats to life and livelihood. It also pursues the transcendent goals of an inclusive democracy, appropriate economic growth, gender equality, and sustainability. The territorial frame of these objectives is the nation-state. Alternative development, therefore, pursues structural changes at the national level as well as local meliorative action. Although political action is necessary to achieve desired structural changes, the alternative approach sees greater access to the bases of social power on the part of households as a step that at least is not inconsistent with political empowerment and may be directly supportive of such a strategy.

It is true, of course, that local actions do not "add up" to national development. While thinking small is important, it can never be enough. Local action of the sort suggested by an alternative development needs to be facilitated, complemented, and supported by appropriate action at the state level.

On hearing this a skeptic might ask why the dominant political elites should voluntarily consent to structural changes that will only benefit the (dis)empowered at the elites' expense. Implicit in this question is the assumption that an alternative development is some kind of zero-sum game. This assumption must be challenged.

In an insightful 1985 article the Chilean economist Sergio Bitar argues

---

[17]   The house journal of the Interamerican Development Foundation, *Grassroots Development*, makes for heartwarming reading of successful local projects involving the poorest of the poor. But the national impact of these projects is virtually nil. It is as if a medical doctor is saving individual lives without affecting the overall mortality rate. Saving individual lives is important, but it is not enough.

that Latin America's prolonged economic crisis is a result of a number of structural characteristics of the countries of the region, including

- A frail national production capacity derived, in part, from the lack of intersectoral integration, especially between agriculture and industry.
- A weak state apparatus with few mechanisms for offsetting cyclical effects, with reduced participation in saving and investment, and without the will or power to regulate relations with the international financial system.
- *A low level of participation and articulation among the main social groups, high marginality and inequality, and consequent deterioration of internal [social] cohesion.*
- The absence of a national endogenous development project, or the lack of popular support for it.
- The crumbling of regional or subregional integration agreements, such as the Andean Trade Pact. (1985, p. 160; my italics).

I have underscored the third of these structural impediments to the historical progress of Latin America because it relates directly to my argument here. If Bitar is correct, one reason for the nearly permanent crisis of Latin American countries is the existence of mass poverty. This is not a spurious form of reasoning but an argument that the economies of Latin America are severely damaged by massive poverty that produces political instabilty and renders difficult, if not impossible, what Bitar calls "the articulation of new and stable social alliances." It is a question of "identifying the social agents [of development], conferring more autonomy and decentralization, and seeking ways to create a new social consensus, or social pact or formulas for concerted action, which will provide a stable basis for supporting a new scheme of development" (1985, p. 160).

If we take this argument seriously, as I am inclined to do – one thinks here of the exemplary income distributions of Japan, South Korea, Hong Kong, Singapore, and Taiwan – then it may be possible to persuade the classes in power to mend their ways, because it will be in their own best interest to do so. A successful politics championing the objectives of an alternative development will have to become the carrier of this message. It will also have to blur the distinctions Aristotelian logic has drawn between market and nonmarket, civil society and the state, an economy of accumulation and one of subsistence, public and private, the individual and group, city and countryside, and self and community while seeking ground between present and future generations, male and female, human beings and the environment. A successful alternative politics would rest on a philosophy that accepts the dialectical character of human existence as a unity of opposites that advances only through conflict and political struggle.

Based on this dialectical view of human progress, advocates of an alternative development would treat mainstream doctrine as only a partial expression of a more inclusive approach to development. Although mainstream doctrine and the alternative model are in conflict, the question is not which is in error (there is, in any case, no ultimate truth) but how each must be modified in practice.

It does not come as a surprise, therefore, that the World Bank has created an office dealing with environmental questions and has begun to assess its projects in environmental terms and even to make loans for environmental education (World Bank, 1989b; Gregersen et al., 1989). Or that it is concerned with such questions as "making the poor creditworthy" (Pulley, 1989). And the World Bank is not alone in its receptivity to the concerns of an alternative development. In one form or another, all international aid agencies, bilateral and multilateral, from UNICEF to the World Health Organization to the United Nations Development Program and Food and Agriculture Organization, have begun to pay attention to questions of gender equality, sustainabilty, and community participation. Although mainstream economic doctrine continues to prevail, it is increasingly being challenged. In truly dialectical fashion, the counter-hegemonic model must work its way into the mainstream and there begin the long process of transforming both the mainstream and itself.

Epilogue: Some Questions for Rich Countries

What I have discussed in these chapters is an approach to a development that is meaningful for the poor in poor countries. But our fates are intertwined. Not only do we in the United States and other countries we call rich have disempowered sectors, but restructuring the economies of these countries may be necessary to promote policies that will also help the poor in poor countries. Some of these policies are now being argued by activists in the environmental movement, but the specific lessons of an alternative development as it might apply to our own disempowered populations remain to be drawn. Rather than do so explicitly and at length, I prefer to pose a series of leading and, I hope, provocative questions that should give cause for reflection.

1 *Are the "dangerous classes" really dangerous?*
The poor, many of whom are found among "Third World" immigrants to the United States, continue to be seen as a threat to law and order. The largest expenditure item in local budgets is for police protection, the judiciary, and jails, – that is, for the repressive apparatus of the state. At the same time, welfare and social services are being scaled back for fear that they might go to "undeserving" (i.e., lazy and unreliable) poor.

But is this the right way to understand poverty? Are the poor poor because of some character deficiency, or are they poor for "structural" reasons? Should the victim be blamed?

In poor countries, the structural character of poverty can be more readily perceived than in rich, where those who do well economically consider that they have morality on their side. But isn't it time to question this convenient assumption? Such questioning has implications for government policy.

2 *What principles should guide the state in its dealings with the excluded minorities?*
If an answer to the first question is that poverty is not a character deficiency but a result of a systematic effort by those in power to keep the poor disempowered, then the current policy of systematic repression is

inappropriate. What is needed instead is the empowerment approach argued in these chapters, with an emphasis on social empowerment.

Citizen rights in the United States are understood largely as political rights, but occasionally there are exceptions. Free education through the twelfth grade has been an established right for generations. By contrast the right to affordable housing, medical care, and child care, to mention only a few possible candidates, remains a politically contested terrain.

The United States prides itself on being a political democracy. But can we claim to being an inclusive democracy as well? Most poor people don't vote. And when they try to organize on their own behalf, often with the help of grass-roots activists, they become the objects of unwanted attention by the country's political police. With social tensions rising, the full weight of government repression comes down on poor people's organizations and the individuals who lead them. But if an empowerment approach is adopted instead, a series of additional questions becomes relevant.

3   *Should an empowerment approach focus on individuals and their competitive ability to move ahead, or on households and their access to the bases of social power?*

The preferred American approach is to help qualified individuals in acquiring the aptitudes and skills necessary to get a job. We tend to glorify individuals who overcome seemingly impossible odds (the one-legged octogenarian who climbs Mount Whitney). We adore winners. Households, on the other hand, are not seen as units worthy of much attention, except perhaps where children are involved (Aid to Families with Dependent Children programs). And with its major allocation of time to domestic (unpaid) tasks, the household economy is rarely taken seriously as a focus of social and economic policy.

By contrast, the empowerment model in chapter 4 is rooted in the existential situation of poor households encapsulated within civil society. What would happen if the household became the starting point for social policy in the United States? Jacqueline Leavitt and Susan Saegert's *Community Households in Harlem* (1989) challenges the traditional emphasis on the individual by looking at how poor people in Harlem act in solidarity with each other when the state isn't watching. When the household is taken as the relevant unit for policy formulation, defensible life space, surplus time, self-organization, and social networking become crucial variables.

4   *What incentives can be devised to encourage communities to organize and undertake initiatives of their own?*

If effective community organization is perceived as a threat to the state, repression follows. On the other hand, activists may decry well-behaved and docile community organizations as Uncle Toms. Perhaps it is necessary to understand state-community interaction as an instance of "antagonistic cooperation."

Some programs have gone beyond this constrained hostility, however, and have arrived at more creative solutions. The so-called solidarity-group loans of Bangladesh's Grameen Bank have been widely imitated (Mann et al., 1989). A genuine participatory approach to development, with communities being involved at all stages of problem identification and in the implementation of meliorative programs, may be another approach. The important thing is to look for inventive solutions that are specific to each case rather than to devise programs that are fully specified before they reach the local community.

5 What are the respective roles of community, state, and private voluntary organizations in a territorially organized process of self-development?

Unlike in poor countries, social policy in the United States has tended to bypass private voluntary organizations (PVOs). Or rather, what has prevailed is an institutional form of specialization. Certain well-established PVOs continue to fund their work chiefly through private donations (churches, the Red Cross, the United Way), while governments devise their own policies and programs. But it is in principle possible to link the NGO/PVO "sector" to policy implementation. Whether or not this is also desirable remains a moot question.

Not all arguments are supportive of this idea. Many small NGOs in the United States do good work on the basis of private grants. The organized local community is yet another collective actor. The political space for collaborative local action involving state, NGOs/PVOs, and organized communities (or popular organizations) needs to be opened up to creative combinations.

6 How can a sense of territory-based identities be strengthened?

In the preceding paragraphs I have used "community" in a loose sense. I had in mind territory-based communities with some degree of self-identity. American ethnic enclaves ("ghettos") may lend themselves to this approach, but their identity image isn't always flattering, because these very same communities are often also the places with the highest incidence of violent and senseless crime. Moreover, many ethnic enclaves have become the dumping ground for the negative "goods" that are unwanted in the more prosperous sections of the city: prisons, freeway interchanges, and toxic-waste incinerators is a small sampling.

The recovery of community pride and identity is not an impossible task, however. Some Asian groups (Japanese, Koreans, Chinese) have managed it well, and we may want to learn from their experience. What are the positive forces in ethnic enclaves that should be strengthened? What can be done to give members of a community a greater voice in matters affecting their life together? Reclaiming territory from violent youth gangs would seem to be a necessary step, but economic opportunities must also be opened up. Where youth unemployment reaches up to 50 percent, large-scale government intervention may be the only feasible answer. But even with government intervention, direct community involvement should be furthered.

7   How can the coproduction of life and livelihood be organized among disempowered households?

To regard households as the ultimate loci – the sacred spaces – of consumption is an ingrained American habit. But an alternative development stresses their productive role instead, and hence turns from privatization toward cooperative models of development. The question, however, doesn't challenge the tradition as such. It merely asks how behavior may be changed to lead to other perceptions. This is similar to question 4, except that it emphasizes household interactions rather than already-organized communities. It asks the question of how neighbors can be induced to coproduce their own lives, emphasizing both market and moral-economy relations. Here the role of local churches might be relevant, because churches bring people together and allow self-organizing activities to happen. The extension of labor unions into communities – which has occurred sporadically in the United States in coal mining areas and during the period of extensive deindustrialization – offers another ray of hope.

8   What planning models are appropriate to household and community self-empowerment?

Several such models may be useful in implementing alternative development. They include both social mobilization and social learning models (Friedmann, 1987). Both require substantial departures from traditional planning practice, which is typically imposed from above rather than generated from within the communities of the disempowered themselves. State agencies wishing to work with empowerment approaches need to become familiar with these models, even if they cannot put them into practice themselves.

9   What constraints in structure and policy must give way to make self-development along an alternative path possible?

The inclusion of the disempowered population, in rich countries will require structural changes in the hegemonic culture. Implied here is a national politics that continues to press for the changes that are needed.

Local changes may involve political redistricting to give minority citizens a greater voice in elections. At the national level, legal battles may have to be fought for key legislation and new citizen rights.

The great battles of the civil rights, feminist, and environmental movements are being fought in this way. Not all of them will be won, and the struggle continues. But it is important to recognize that local community struggles and political action at higher territorial levels, far from being incompatible, are mutually supportive and complementary. Moreover, positive contributions are made by all factions, from mainstream organizations to the strident radical fringe.

Alternative development must become a global project.

Bibliography

Abdullah, Tahrunnessa. 1980. Women's participation in rural development: a
  Bangladesh pilot project. In International Labour Office, *Women in rural
  development: critical issues.* Geneva: ILO.
Åckerman, Nordal. 1979. Can Sweden be shrunk? *Development Dialogue,* no. 2,
  71–114.
Adams, Richard Newbold. 1988. *The eighth day: social evolution as the self-
  organization of energy.* Austin: University of Texas Press.
Agarwal, Arril; D'Monte, Darryl; and Samarth, Ujwala, eds. 1987. *The fight for
  survival: people's action for environment.* New Delhi: Centre for Science and
  Environment.
Ahluwalia, M. S. 1974. Income inequality: some dimensions of the problem. In
  Hollis Chenery et al., *Redistribution with growth.* Published for the World
  Bank and the Institute of Development Studies, Sussex. New York: Oxford
  University Press.
Amnesty International. 1988. *The Universal declaration of human rights, 1948–
  1988.* Special edition. New York: Amnesty International USA.
Andreas, Carol. 1985. *When women rebel: the rise of popular feminism in Peru.*
  Westport, Conn.: Lawrence Hill.
Andrus, Beth. 1988. The universal declaration of human rights. In *The universal
  declaration of human rights, 1948–1988,* pp. 1–16. New York: Amnesty
  International USA.
Annis, Sheldon, and Hakim, Peter, eds. 1988. *Direct to the poor: grassroots
  development in Latin America.* Boulder, Colo., and London: Lynne Rienner
  Publishers.
Ardaya, Gloria. 1986. Trabajadores informales en La Paz: el caso de las vende-
  doras ambulantes. In *El sector informal en Bolívia.* La Paz: FLACSO, CEDLA.
Arendt, Hannah. 1958. *The human condition.* Chicago: University of Chicago
  Press.
Arndt, Heinz Wolfgang. 1987. *Economic development: the history of an idea.*
  Chicago: University of Chicago Press.
Ayres, Robert L. 1983. *Banking on the poor: the World Bank and world poverty.*
  Cambridge, Mass.: MIT Press.
Badri, Balghis, 1990. A profile of Sudanese women. In *Institutional and sociopo-
  litical issues.* Vol. 3 of *The long-term perspective study of Sub-Saharan Africa.*
  Washington, D.C.: World Bank.

Ballón, Eduardo, ed. 1986. *Movimientos sociales y democrácia: la fundación de un nuevo orden.* Lima: DESCO.

Ballón, Eduardo. 1989. La planificación participativa y la organización comunitaria como via del desarrollo. El caso de Villa El Salvador, Lima, Perú. Lima: DESCO.

Banfield, Edward C. 1970. *The unheavenly city: the nature and future of our urban crisis.* Boston: Little Brown.

Bangladesh Rural Advancement Committee 1983. Household strategies in Bonkura Village. In David C. Korten, ed., 1986, *Community management: Asian experiences and perspectives.* West Hartford, Conn.: Kumarian Press.

Baran, Paul A., and Sweezy, Paul M. 1966. *Monopoly capital: an essay on the American economic and social order.* New York: Monthly Review Press.

Barrig, Maruja. 1989. The difficult equilibrium between bread and roses: women's organizations and the transition from dictatorship to democracy in Peru. In Jane S. Jaquette, ed., *The women's movement in Latin America: feminism and the transition to democracy,* chapter 5. Boston: Unwin Hyman.

Barrios Villegas, Franz. 1987. ONG's y realidad nacional. In UNITAS, *El rol de las ONG's en Bolívia,* pp. 3–8. La Paz: UNITAS.

Beatley, Timothy. 1989. Environmental ethics and planning theory. *Journal of Planning Literature,* 4, no. 1, 1–32.

Bell, C. L. G. 1974. The political framework. In Hollis Chenery et al., *Redistribution with growth.* Published for the World Bank and the Institute of Development Studies, Sussex. New York: Oxford University Press.

Benería, Lourdes. 1980. Some questions about the origin of the division of labour by sex in rural societies. In International Labour Office, *Women in rural development: critical issues.* Geneva: ILO.

Berger, Peter L., and Neuhaus, Richard John. 1977. *To empower people: the role of mediating structures in public policy.* Washington, D.C.: American Enterprise Institute for Public Policy Research.

Bhatt, Chandi Prasat. 1990. The Chipko Andolan: forest conservation based on people's power. *Environment and Urbanization,* 2, no. 1, 7–18.

Biggs, Tyler; Grindle, Merilee S.; and Snodgrass, Donald R. 1988. The informal sector, policy reform, and structural transformation. In Jerry Jenkins, ed., *Beyond the informal sector: including the excluded in developing countries.* A Sequoia Seminar. San Francisco: ICS Press.

Bitar, Sergio. 1985. The nature of the Latin American crisis. *CEPAL Review,* no. 27, 159–64.

Block, Fred L. 1990. *Postindustrial possibilities: a critique of economic discourse.* Berkeley: University of California Press.

Borja, Jordi, et al., eds. 1987. *Descentralización del estado o movimiento social y gestión local.* Santiago: FLACSO, CLACSO, ICI.

Braybrooke, David. 1987. *Meeting needs.* Princeton, N.J.: Princeton University Press.

Breman, Jan. 1985. A dualistic labour system? A critique of the "informal sector" concept. In Ray Bromley, ed., *Planning for small enterprises in Third World cities.* New York: Pergamon Press.

Browner, C. H. 1986. Gender roles and social change: a Mexican case study. *Ethnology*, 25, no. 2, 89–106.

Bunster, Ximena, and Chaney, Elsa M. 1985. *Sellers and servants: working women in Lima, Peru.* New York: Praeger.

Burns, Leland S. 1970. *Housing: symbol and shelter.* With Robert G. Healy, Donald McAllister, and B. Khing Tjioe. Los Angeles: International Housing Productivity Study, Graduate School of Business Administration, UCLA.

Bury, John B. 1920. *The idea of progress. An inquiry into its origins and growth.* London: Macmillan.

Calderón, Fernando, ed. 1986. *Los movimientos sociales ante la crisis.* Buenos Aires: CLACSO.

Calderón Cockburn, Julio, and Olivera Cárdenas, Luis. 1989. *Municipio y pobladores en la habilitación urbana (Huaycán y Laderas del Chillón).* Lima: DESCO.

Campero, Guillermo. 1987. *Entre la sobrevivencia y la acción política: las organizaciones de pobladores en Santiago.* Santiago; Estudios ILET.

Carr, Marilyn. 1981. *Developing small-scale industries in India: an integrated approach.* London: Intermediate Technology Publications.

Carr, Marilyn. ed. 1985. *The AT reader: theory and practice in appropriate technology.* New York: Intermediate Technology Group of North America.

Castells, Manuel. 1983. *The city and the grassroots.* Berkeley: University of California Press.

Castells, Manuel, and Laserna, Roberto. 1989. La nueva dependencia: Cambio tecnológico y reestructuración socioeconómica en América Latina. *David y Goliath.* Revista del Consejo Latinoamericano de Ciencias Sociales, 18, no. 55, 2–16.

Chambers, Robert. 1987. *Sustainable rural livelihoods: a strategy for people, environment, and development.* An overview paper for Only One Earth: Conference on Sustainable Development. London: International Institute for Environment and Development.

———. 1983. *Rural development: putting the last first.* London: Longman.

Chaney, Elsa M., and Castro, Mary García, eds. 1989. *Muchachas no more: household workers in Latin America and the Caribbean.* Philadelphia: Temple University Press.

Chávez O'Brien, Eliana. 1988. *El sector informal urbano: de reproducción de la fuerza de trabajo a posibilidades de producción.* Lima: Fundación Friedrich Ebert.

Cloud, Kathleen. 1985. Women's productivity in agricultural systems: considerations for project design. In Catherine Overholt et al., eds., *Gender roles in development projects: a case book.* West Hartford, Conn.: Kumarian Press.

Cockburn, Cynthia. 1977. *The local state: management of cities and people.* London: Pluto Press.

"The Cocoyoc Declaration." 1974. *Development Dialogue*, no. 2, 88–96. Uppsala, Sweden: Dag Hammarskjöld Foundation.

Colby, Michael E. 1990. *Environmental management in development: the evolution of paradigms.* Discussion Paper 80. Washington, D.C.: World Bank.

Colombo, Daniela; Frey, Luigi; and Livraghi, Renata. 1988. The response of public authorities in Italy to the needs expressed by women. In Kate Young, ed., *Women and economic development: local, regional and national planning strategies*, chapter 2. New York: Berg Publishers; Paris: UNESCO.

CSUTCB (Confederación Sindical Unica de Trabajadores Campesinos de Bolívia). 1989. *Documentos y resoluciones del I Congreso Extraordinario de la C.S.U.T.C.B. 1 al 17 de julio, 1988. Potosí.*

Coraggio, José Luis. 1988. Poder local, poder popular? To be published in *Cuadernos de CLAEH*, Montevideo.

Cotler, Julio, ed. 1987. *Para afirmar la democrácia*. Lima: IEP.

Crozier, Michel; Huntington, Samuel P.; and Watanuki, Joji. 1975. *The crisis of democracy: report on the governability of democracies to the trilateral commission*. New York: New York University Press.

Dag Hammarskjöld Foundation. 1975. What now? Another development. Report prepared on the occasion of the Seventh Special Session of the United Nations General Assembly. Special issue of *Development Dialogue*, nos. 1–2.

Dahl, Robert A. 1990. *Democracy and its critics*. New Haven, Conn.: Yale University Press.

Daly, Herman E. ed. 1980. *Economics, ecology, ethics: essays towards a steady-state economy*. San Francisco: W. H. Freeman & Co.

Das, Amritnanda. 1979. *Foundations of Ghandian economics*. Bombay: Allied Press.

de Neufville, Judith Innes. 1984. Applicability of social indicators in the United States of America: a century of social policy and social indicators. In UNESCO, *Applicability of indicators of socio-economic change for development planning*. Paris: UNESCO.

de Soto, Hernando. 1989. *The other path: the invisible revolution in the Third World*. New York: Harper & Row.

Deval, Bill, and Sessions, George. 1985. *Deep ecology: living as if nature mattered*. Salt Lake City: Peregrine Smith Books.

Douglass, Michael. 1985. The regional impact of transmigration. Report to the World Bank. Honolulu: University of Hawaii. Manuscript.

Downs, Anthony. 1967. *Inside bureaucracy*. Boston: Little Brown.

Drabek, Anne Gordon, ed. 1987. *Development alternatives: the challenge for NGOs*. Supplement to *World Development*, vol. 15.

Drewnowski, Jan. 1974. *On measuring and planning the quality of life*. The Hague: Mouton.

Duchacek, Ivan. 1970. *Comparative federalism: the territorial dimension of politics*. New York: Holt, Rinehart & Winston.

Eckstein, Susan. 1988. *The poverty of revolution: the state and the urban poor in Mexico*. 2d ed. Princeton; N.J.: Princeton University Press.

Edmundson, W. C., and Sukhatme, P. V. 1990. Food and work: poverty and hunger? *Economic Development and Cultural Change*, 38, no. 2, 263–80.

Ekeh, Peter P. 1974. *Social exchange theory: the two traditions*. Cambridge, Mass.: Harvard University Press.

Elkins, Paul, ed. 1986. *The living economy: a new economics in the making*. New York: Routledge.

Environment and Urbanization 1989. 1, no. 1.

——. 1990. "SPARC: developing new NGO lines." 2, no. 1, 91–104.

Etzioni, Amitai. 1988. *The moral dimension: toward a new economics.* New York: The Free Press.

Evans, Peter, and Stephens, John D. 1988. "Studying development since the sixties: the emergence of a new comparative political economy." *Theory and Society*, 17, 713–45.

Evers, Hans-Dieter. 1989. Urban poverty and labor-supply strategies in Jakarta. In Gerry Rodgers, ed., *Urban poverty and the labor market*. Geneva: ILO.

Evers, Tilman. 1985. Identity: the hidden side of new social movements in Latin America. In David Slater, ed., *New social movements and the state in Latin America*. Amsterdam: Centre for Latin American Research and Documentation.

Falk, Richard. 1981. *Human rights and state sovereignty*. New York: Holmes & Meier.

Falk, Richard; Kim, Samuel S.; and Mendlovitz, Saul H., eds. 1982. *Toward a just world order*. Vol. 1 of *Studies on a just world order*. Boulder, Colo.: Westview Press.

Fals Borda, Orlando. 1985. *El problema de como investigar la realidad para transformarla por la praxis*. 3d rev. ed. Bogotá: Ediciones Tercer Mundo.

——. 1986. *Conocimiento y poder popular: lecciones con campesinos de Nicaragua, México y Colombia*. Bogotá: Punta de Lanza, Siglo Veintiuno Editores.

Fass, Simon M. 1988. *Political economy in Haiti: the drama of survival*. New Brunswick, N.J.: Transaction Books.

Fass, Simon. 1989. *Street food vendors in the industrial area of Port-au-Prince*. Port-au-Prince, Haiti: Centre de Promotion des Femmes Ouvrières.

Feijoó, Maria del Carmen, and Gogna, Monica, 1985. Las mujeres en la transición a la democrácia. In Elizabeth Jelín, ed., *Los nuevos movimientos sociales*. Vol. 1, *Mujeres. Rock nacional*, chapter 2. Buenos Aires: Centro Editor de América Latina.

Financing ecological destruction. 1987. Material prepared for presentation at the World Bank/IMF meeting, September 29 to October 1, 1987. Endorsed by 28 international nongovernmental organizations "concerned with the preservation of Tropical Forests/Wetlands and the indigenous peoples who live within them."

Fort, Amelia. 1988. La mujer en la política de servícios. In Maruja Barrig, ed., *De vecinas a ciudadanas: la mujer e en el desarrollo urbano*. Lima: SUMBI.

Freire, Paulo. 1973. *Education for critical consciousness*. New York: Seabury Press.

——. 1981. *Pedagogy of the oppressed*. New York: Continuum.

Friedmann, Harriet. 1980. Household production and the national economy: concepts for the analysis of agrarian formations. *Journal of Peasant Studies*, 7, no. 2, 158–84.

Friedmann, John. 1966. *Regional development policy: a case study of Venezuela*. Cambridge, Mass.: MIT Press.

——. 1968. A strategy of deliberate urbanization. *Journal of the American Institute of Planners*, 34, no. 6, 364–73.

——. 1979. Basic needs, agropolitan development, and planning from below. *World Development*, 7, no. 6, 607–13.

——. 1987. *Planning in the public domain: from knowledge to action*. Princeton, N.J.: Princeton University Press.

——. 1988. *Life space and economic space: essays in Third World planning*. Brunswick, N.J.: Transaction Books.

——. 1989a. The dialectic of reason. *International Journal of Urban and Regional Research*, 13, 217–33.

——. 1989b. The Latin American *barrio* movement as a social movement: contribution to a debate. *International Journal of Urban and Regional Research*, 13, no. 3, 501–10.

Friedmann, John, and Forest, Yvon. 1988. The politics of place: toward a political economy of territorial planning. In Benjamin Higgins and Donald J. Savoie, eds., *Regional economic development: essays in honour of François Perroux*. Boston: Unwin Hyman.

Friedmann, John, and Weaver, Clyde. 1979. *Territory and function: the evolution of regional planning*. London: Edward Arnold.

Galín, Pedro; Carrión, Julio; and Castillo, Oscar. 1986. *Asalariados y clases populares en Lima*. Lima: Instituto de Estudios Peruanos.

Garilao, Ernesto D. 1987. Indigenous NGOs as strategic institutions: managing the relationship with government and resource agencies. *World Development*, 15, supplement, 113–20.

Ghai, Dharam P. 1977. What is the basic needs approach to development all about? In International Labour Office, *The basic-needs approach to development: some issues regarding concepts and methodology*. Geneva: ILO.

Ghai, Dharam; Hopkins, Michael; and McGranahan, Donald. 1988. *Some reflections on human and social indicators for development*. Discussion Paper No. 6. Geneva: United Nations Research Institute for Social Development.

Gilligan, Carol; Ward, Janie Victoria; and Taylor, Jill McLean, eds. 1988. *Mapping the moral domain*. Cambridge, Mass.: Harvard University Press.

Glaeser, Bernard. ed. 1984. *Eco-development: concepts, policies, strategies*. New York: Pergamon Press.

Gregersen, Hans; Draper, Sydney; and Elz, Dieter. 1989. *People and trees: the role of social forestry in sustainable development*. EDI Seminar Series. Washington, D.C.: World Bank.

Griffin, Keith. 1989. Alternative strategies of development. Los Angeles: University of California, Center for Social Theory and Conparative History, Colloquium Series. Unpublished paper.

Grillo, Oscar Jorge. 1988. *Articulación entre sectores urbanos populares y el estado local. El caso del barrio de la Boca*. Buenos Aires: Centro Editor de América Latina.

Grindle, Merilee S. 1988. *Searching for rural development, labor migration and employment in Mexico*. Ithaca, N.Y.: Cornell University Press.

Gross, Bertram M. 1966. *The state of the nation: social systems accounting*. London: Tavistock Publications.

Gurley, John G. 1976 *China's economy and the Maoist strategy*. New York: Monthly Review Press.

Guyer, Jane I., and Peters, Pauline E., eds. 1987. Conceptualizing the household: issues of theory and policy in Africa. Special issue. *Development and Change*, 18, no. 2.

Hall, Peter. 1982. *Great planning disasters*. Berkeley: University of California Press.

———. 1988. *Cities of tomorrow*. London: Bart Blackwell.

———. 1990. International urban systems. Working Paper 514. Berkeley: University of California, Institute of Urban and Regional Development.

Hardin, Garrett. 1980. The tragedy of the commons. In Herman E. Daly, ed., *Economics, ecology, ethics: essays towards a steady-state economy*, chapter 6. San Francisco: W. H. Freeman & Co.

Hardoy, Jorge E., and Satterthwaite, David. 1989. *Squatter citizen: life in the urban world*. London: Earthscan Publications.

Hardy, Clarisa. 1984. *Los talleres artesanales de Conchalí: la organización, su recorrido y sus protagonistas*. Santiago: Academía de Humanismo Cristiano (PET).

———. 1986. *Hambre + dignidad = ollas comunes*. Santiago: Academía de Humanismo Cristiano (PET).

Harris, Olivia. 1981. Households as natural units. In Kate Young, Carol Wolkowitz, and Roslyn McCullagh, eds., *Of marriage and the market*. London: Routledge & Kegan Paul.

Hecht, Susanna, and Cockburn, Alexander. 1989. *The fate of the forest: developers, destroyers and defenders of the Amazon*. London and New York: Verso.

Heller, Agnes, 1976. *The theory of needs in Marx*. London: Allison & Busby.

Herzer, Hilda, and Pírez, Pedro, eds. 1988. *Gobierno de la ciudad y crisis en Argentina*. Buenos Aires: IIED-América Latina and Grupo Editor Latinoamericano.

Higgins, Benjamin. 1976. The unified approach to development planning at the regional level: the case of Pahang Tenggara. In Antoni Kuklinski, ed., *Regional development in worldwide perspective*. The Hague: Mouton.

Hirschman, Albert O. 1981. The rise and decline of development economics. In *Essays in trespassing*. Cambridge: Cambridge University Press.

———. 1984. *Getting ahead collectively: grassroots experiences in Latin America*. New York: Pergamon Press.

Honadle, George, and VanSant, Jerry. 1985. *Implementation for sustainability: lessons from integrated rural development*. West Hartford, Conn.: Kumarian Press.

Hueting, Roefie. 1980. *New scarcity and economic growth: more welfare through less production?* Amsterdam: North Holland Publishing Co.

Hunt, Diana. 1984. *The impending crisis in Kenya: the case for land reform*. London: Gower.

———. 1989. *Economic theories of development*. New York: Harvester Wheatsheaf.

Hyden, Goran. 1980. *Beyond Ujamaa in Tanzania: underdevelopment and an uncaptured peasantry*. Berkeley: University of California Press.

Iglesias, Enrique V. 1989. A social policy without paternalism. *Grassroots Development* (journal of the Interamerican Foundation), 13, no. 1, 41–2.

Iliffe, John. 1987. *The African poor: a history.* Cambridge: Cambridge University Press.

Interamerican Foundation. 1990. *A guide to NGO directories.* Rosslyn, Va.: Interamerican Foundation.

International Foundation for Development Alternatives. 1980. *Dossier*, no. 17.

International Labour Office. 1976a. *Tripartite world conference on employment, income distribution, and social progress and the international division of labour: background papers.* Vol. 1, *Basic needs and national employment strategies.* Geneva: ILO.

———. 1976b. *Employment, growth and basic needs.* Tripartite World Conference on Employment, Income Distribution, and Social Progress and the International Division of Labour. Report of the director general. Geneva: ILO.

———. 1977. *Meeting basic needs: strategies for eradicating mass poverty and unemployment.* Geneva: ILO.

Jacobi, Pedro. 1989. *Movimentos sociais e politicas públicas: demandas por saneamento básico e saúde.* São Paulo, 1974–84. São Paulo: Cortez Editorial.

Jain, Devaki. 1989. Letting the worm turn: a comment on innovative poverty alleviation. In William P. Lineberry, ed., *Assessing participatory development: rhetoric versus reality.* Boulder, Cool.: Westview Press.

Jaquette, Jane S., ed. 1989. *The women's movement in Latin America: feminism and the transition to democracy.* Boston: Unwin Hyman.

Korten, David C. 1980. Community organization and rural development: a learning process approach. *Public Administration Review*, 40, 480–510.

Korten, David C., ed. 1986. *Community management: Asian experience and perspectives.* West Hartford, Conn.: Kumarian Press.

———. 1987. Third generation NGO strategies: a key to people-centered development. *World Development*, 15, supplement, 145–59.

———. 1990. *Getting to the 21st century: voluntary development action and the global agenda.* West Hartford, Conn.: Kumarian Press.

Korten, David C., and Klaus, Rudi, eds. 1984. *People-centered development: contributions toward theory and planning frameworks.* West Hartford, Conn.: Kumarian Press.

Kowarich, Lucio, ed. 1988. *As lutas sociais e a cidade de São Paulo: passado e presente.* São Paulo: Editora Paz e Terra.

Kropotkin, Peter. 1970. *Selected writings on anarchism and revolution.* Edited by Martin A. Milley. Cambridge, Mass.: MIT Press.

Landim, Leilah. 1987. Non-governmental organization in Latin America. *World Development*, 15, supplement, 29–38.

Leavitt, Jacqueline, and Saegert, Susan. 1989. *From abandonment to hope: Community households in Harlem.* New York: Columbia University Press.

Lechner, Norbert. 1988. *Los patios interiores de la democrácia.* Santiago: FLACSO.

Leckie, Scott. 1989. Housing as a human right. *Environment and Urbanization*, 1, no. 2, 90–108.

Lee, E. L. H. 1977. Some normative aspects of a basic needs strategy. In International Labour Office, *The basic-needs approach to development: some issues regarding concepts and methodology.* Geneva: ILO.

Leiss, William. 1976. *The limits to satisfaction: an essay on the problem of needs and commodities*. Toronto: University of Toronto Press.

Lindblom, Charles. 1977. *Politics and markets*. New York: Basic Books.

Lindholm, Stig. 1976. "Another Sweden:" how the Swedish press reacted. *Development Dialogue*, no. 1, 68–82.

Little, Ian M. D.; Mazumdar, Dipak; and Page, John M., Jr. 1987. *Small manufacturing enterprises: a comparative study of India and other economies*. Published for the World Bank. New York: Oxford University Press.

Logan, Kathleen. 1989. Latin American urban mobilizations: women's participation and self-empowerment. In Gmelch and Zenner, eds., *Urban Life: Readings in Urban anthropology*. New York: St. Martin's Press. 2d ed.

Lomnitz, Larissa Adler. 1977. *Networks and marginality: life in a Mexican shantytown*. New York: Academic Press, 1977.

McCall, Michael, and Skutsch, Margaret. 1983. Strategies and contradictions in Tanzania'a rural development: which path for the peasants? in David A. M. Lea and D. P. Chaudhri, eds., *Rural development and the state*. London: Methuen.

McDonald, Mark. 1989. Dams, displacement, and development in Brazil: a case study of the Uruguai River Basin Project, 1979–1989. Graduate School of Architecture and Urban Planning. Los Angeles: University of California, M. A. thesis.

McGranahan, D. V. et al., 1972. *Contents and measurement of socioeconomic development*. A Staff Study of the United Nations Research Institute for Social Development. New York: Praeger.

McGranahan, Donald, Pizarro, Eduardo; and Richard, Claud. 1985. *Measurement and analysis of socioeconomic development – an enquiry into international indicators of development and quantitative interrelations of social and economic components of development*. Geneva: United Nations Research Institute for Social Development.

Machado, Leda M. V. 1987. The problems for women-headed households in a low-income housing programme in Brazil. In Caroline O. M. Moser and Linda Peake, eds., *Women, human settlements, and housing*, chapter 3. London: Tavistock Publications.

McNamara, Robert S. 1973. *Address to the board of governors* [Nairobi, Kenya]. Washington, D.C.: World Bank.

Mangahas, Mahar. 1977. The Philippines social indicators project. *Social Indicators Research*, 4, 67–96.

Mann, Charles K.; Grindle, Merilee S.; and Shipton, Parker, eds. 1989. *Seeking solutions: framework and cases for small enterprise development programs*. West Hartford, Conn.: Kumarian Press.

Manuel, F. E. and Manuel, F. P. 1979. *Utopian thought in the Western world*. Oxford: Basil Blackwell.

Markusen, Ann. 1987. *Regions: the economics and politics of territory*. Totowa, N.J.: Rowman & Littlefield.

Martens, Bertin. 1989. *Economic development that lasts: labour-intensive irrigation projects in Nepal and the United Republic of Tanzania*. Geneva: ILO.

Martinez-Alier, Juan. 1987. *Ecological economics: energy, environment, and society.* Oxford: Basil Blackwell.

Mathew, N. T. and Scott, Wolf. 1985. *A development monitoring service at the local level.* Geneva: United Nations Research Institute for Social Development.

Matos Mar, José. 1985. *Desborde popular y crisis del Estado: el nuevo rostro del Perú en la década de 1980.* 2d ed. Lima: Instituto de Estudios Peruanos.

Meadows, Donella H., et al., 1972. *The limits to growth: a report of the Club of Rome's project on the predicament of mankind.* New York: Universe Books.

Meier, Gerald M. and Seers, Dudley, eds. 1984. *Pioneers in development.* Published for the World Bank. New York: Oxford University Press.

Meier, Richard L. 1980. A stable urban ecosystem: anticipations for the Third World. *Third World Planning Review,* 2, no. 2, 2, 153–69.

Meier, R. L.; Berman, Sam; Campbell, Tim; and Fitzgerald, Chris; 1981. *The urban ecosystem and resource-conserving urbanisms in Third World cities.* Berkeley: University of California, Lawrence Berkeley Laboratory, LBL–12640.

Miles, Ian. 1985. *Social indicators for human development.* London: Francis Pinter.

Moctezuma, Pedro. 1990. Mexico's urban popular movements. *Environment and Urbanization,* 2, no. 1, 35–50.

Moisés, José Alvaro. 1981. O estado, as contradições urbanas, e os movimentos sociais. In J. A. Moisés, ed., *Cidade, povo e poder.* São Paulo: CEDEC, Paz e Terra.

———. 1986. Sociedade civil, cultura politica e democracia: descaminhos de transição politica. In Maria de Lourdes Manzini Covre, ed., *A cidadania que não tenemos.* São Paulo: Brasiliense.

Molyneux, Maxine. 1985. Mobilisation without emancipation? Women's interests, the state, and revolution in Nicaragua. *Feminist Studies,* 11, no. 2, 227–54.

Morgan, Mary. 1990. Stretching the development dollar: the potential for scaling-up. *Grassroots Development,* 14, no. 1, 2–11.

Moser, Caroline O. N. 1989. Gender planning in the Third World: meeting practical and strategic gender needs. *World Development,* 17, no. 11, 1799–1825.

Moser, Caroline O. N. and Peake, Linda, eds. 1987. *Women and human settlements and housing.* London: Tavistock Publications.

Mumford, Lewis. 1938. *The culture of cities.* New York: Harcourt, Brace & Co.

Mumtas, Khawar, and Shaheed, Farida, eds. 1987. *Women of Pakistan: two steps forward, one step back?* London: Zed Books, Ltd.

Nash, June, 1990. Latin American women in the world capitalist crisis. *Gender and Society,* 4, no. 3, 338–353.

National Perspectives Quarterly 1990. 7, no. 1.

Netherlands Ministry of Housing, Physical Planning and Environment. 1989. *To choose or to lose: national environmental policy plan.* The Hague: SDU Publishers.

Newell, Kenneth W., ed. 1975. *Health by the people.* Geneva: World Health Organization.

Ngau, Peter M. 1989. Rural-urban relations and agrarian development in Kutus area, Kenya. Los Angeles: Urban Planning Program, University of California. Ph.D. dissertation.

Nicholls, William M., and Dyson, William A. 1983. *The informal economy: where people are the bottom line.* Ottawa: Vanier Institute of the Family.

Nielson, John A., and Kehoe, Timothy G. 1987. Housing in Colombia: the role of the RAIC. Royal Architectural Institute of Canada. *Update,* 10, no. 1, 2–5.

Nisbet, Robert A. 1953. *The quest for community: a study in the ethics of order and freedom.* New York: Oxford University Press.

Nisbet, Robert A., and Perrin, Robert. 1977. *The social bond.* 2d ed. New York: Knopf.

Nordhaus, W., and Tobin, J. 1972. Is growth obsolete? In F. T. Juster, ed., *Economic growth: fiftieth anniversary colloquium V.* New York: National Bureau of Economic Research.

Norgaard, Richard B. 1988. Sustainable development: a co-evolutionary view. *Futures,* 20, no. 2, 606–20.

———. 1989a. The case for methodological pluralism. *Ecological Economics,* 1, 35–37.

———. 1989b. Three dilemmas of environmental accounting. *Ecological Economics,* 1, 303–14.

Novak, Michael, ed. 1980. *Democracy and mediating structures.* Washington, D.C.: American Enterprise Institute for Public Policy Research.

O'Donnell, Guillermo, and Schmitter, Phillippe C. 1986. *Transitions from authoritariam rule: tentative conclusions about uncertain democracies.* Baltimore: Johns Hopkins University Press.

Okin, Susan Moller. 1981. Liberty and welfare: some issues in human rights theory. In J. Roland Pennock and John W. Chapman, eds., *Human rights,* chapter 12. New York: New York University Press.

Olson, Mancur. 1965. *The logic of collective action: public goods and the theory of groups.* Cambridge, Mass.: Harvard University Press.

Paauw, Douglas S., and Fei, John C. H. 1973. *The transition in open dualistic economies: theory and Southeast Asian experience.* New Haven Conn.: Yale University Press.

Pahl, R. E. 1989. Is the emperor naked? Some questions on the adequacy of sociological theory in urban and regional research. *International Journal of Urban Regional Research,* 13, no. 4, 709–20.

Palma, Diego. 1987. *La informalidad, lo popular y el cambio social.* Lima: DESCO.

———. 1988. Presupuestos teóricos de la promoción. In Mariano Castro and Enrique Quedena, eds., *Derecho, promoción social y sectores populares y urbanos.* Lima: DESCO.

Papanek, Hanna. 1990. From each less than she needs, from each more than she can do: allocations, entitlements, and value. In Irene Tinker, ed., *Persistent inequalities: women and world development.* New York: Oxford University Press.

Pease García, Henry. 1988. *Democrácia local: reflexiones y experiencias.* Lima: DESCO.

Peattie, Lisa, and Rein, Martin. 1983. *Women's claims: a study in political economy.* New York: Oxford University Press.

Perloff, Harvey S. 1985. Relative regional economic growth: an approach to regional accounts. In Leland S. Burns and John Friedmann, eds., *The art of planning: selected essays of Harvey S. Perloff,* chapter 13. New York: Plenum Press.

Pestel, Eduard. 1989. *Beyond the Limits to Growth: A Report to the Club of Rome.* New York: Universe Books.

Pezzoli, Keith. 1990. The politics of land allocation in Mexico City's ecological zone: the case of Ajusco. Los Angeles: University of California, Urban Planning Program. Ph.D. dissertation.

Pipping, Hugo E. 1953. *Standard of living: the concept and its place in economics.* Vol. 8, no. 4, Helsingörs, Denmark: Societas Scientiarium Fennica Comentationes Humanarum Litterarum.

Piven, Frances Fox, and Cloward, Richard A. 1979. *Poor people's movements: why they succeed, how they fail.* New York: Vintage Books.

Polanyi, Karl. 1977. *The livelihood of man.* Edited by Harry W. Pearson. New York: Academic Press.

Pulley, Robert V. 1989. *Making the poor creditworthy: a case study of the integrated rural development program in India.* World Bank Discussion Paper No. 58. Washington, D.C.: World Bank.

Quijano, Aníbal. 1988. Otra noción de lo privado, otra noción de lo público: notas para un debate latinoamericano. *Revista de la CEPAL,* no. 35, 101–15.

Raczynski, Dagmar, and Serrano, Claudia. 1985. *Vivir la pobreza: testimonios de mujeres.* Santiago. CIEPLAN.

Radin, Margaret Jane. Radin 1987. Market inalienability. *Harvard Law Review,* 100, no. 8, 1848–1937.

Ramzi, Sonia Abadir, and the Centre for Social Science Research and Documentation for the Arab Region. 1988. Women and development planning: the case of Egypt. In Kate Young, ed., *Women and economic development.* New York: Berg Publishers; Paris: UNESCO.

RECEM. 1986. *Municípios em busca de soluções.* São Paulo: Fundação Prefeito Faría Lima-CEPAM.

Redclift, Michael. 1987. *Sustainable development: exploring the contradictions.* New York: Methuen.

——. 1989. The environmental consequences of Latin America's agricultural development: some thoughts on the Brundtland Commission Report. *World Development,* 17, no. 3, 357–63.

Reid, Walter V.; Barnes, James N.; and Blackwelder, Brent. 1988. *Bankrolling successes: a portfolio of sustainable projects.* Washington, D.C.: Environmental Policy Institute and National Wildlife Federation.

Richards, Paul. 1985. *Indigenous agricultural revolution: ecology and food production in West Africa.* London: Hutchinson.

Riofrio, Gustavo, and Dirant, J. C. 1989. *¿Qué vivienda han construido? Nuevos problemas en viejas barriadas.* Lima: CIDAP, TAREA.

Rivas, Alfonso. 1989. Política en el movimiento urbano de la Ciudad de México:

el caso de la Asamblea de Barrios. Los Angeles: University of California, Urban Planning Program. M. A. thesis.

Rodgers, Gerry. ed. 1989. *Urban poverty and labour markets: access to jobs and incomes in Asian and Latin American cities.* Geneva: ILO.

Rodriguez, Alfredo. 1983. *Por una ciudad democrática.* Santiago: SUR.

Rodwin, Lloyd, ed. 1969. *Planning urban growth and regional development: the experience of the Guayana program in Venezuela.* Cambridge, Mass.: MIT Press

Rojas Julca, Julio Andrés. 1989. *Gobierno municipal y participación ciudadana: experiencias de Lima metropolitana, 1984–1986.* Lima: Fundación Friedrich Ebert.

Ross, David P., and Usher, Peter J. 1986. *From the roots up: economic development as if community mattered.* Croton-on-Hudson, N.Y.: Bootstrap Press.

Ruddick, Sara. 1989. *Maternal thinking: towards a politics of peace.* New York: Ballantine Books.

Ruttan, Vernon W. 1975. Integrated rural development programs: a skeptical perspective. *International Development Review,* 17, no. 4, 9–16.

Sabatini, Franciso. 1989. Participación de pobladores en organizaciones de barrio. *EURE,* 15, no. 46, 47–68.

Sachs, Ignacy, 1988. Market, non-market, and the "real" economy. In Kenneth Arrow, ed., *Basic issues,* pp. 218–31. Vol. 1 of *The balance between industry and agriculture in economic development.* London: Macmillan.

Sandoval, Godofredo Z. 1988. *Organizaciones no gubernamentales de desarrollo en América Latina y el Caribe.* 2d ed. La Paz: CEBEMO-UNITAS.

Santana, Pedro. 1983. *Desarrollo regional y paros cívicos en Colombia.* Bogotá: CINEP.

Sanyal, Bishwarpriya. Forthcoming. Antagonistic cooperation: a case study of non-governmental organizations, government and donors' relationships in income generating projects in Bangladesh. *World Development.*

Sasaki, Hideyuki. 1989. Indonesia's transmigration program and regional development on the frontier of East Kalimantan. Los Angeles: University of California. Manuscript.

Schejtman, Alejandro. 1983. Análisis integral del problema alimentario y nutricional en América Latina. *Estudios rurales latinoamericanos,* 6, 2–3.

Schmink, Marianne. 1984. Household economic strategies: review and research agenda. *Latin American Research Review,* 19, no. 3, 87–102.

Schumpeter, Joseph. 1942. *Capitalism, socialism, and democracy.* New York: Harper & Row.

Schwartzman, Stephan. 1986. *Bankrolling disasters: international development bank and the global environment. A citizen's environmental guide to the world.* San Francisco: Sierra Club.

Scott, James C. 1976. *The moral economy of the peasant: rebellion and subsistence in Southeast Asia.* New Haven Conn.: Yale University Press.

———.1985. *Weapons of the weak: everyday forms of peasant resistance.* New Haven Conn.: Yale University Press.

———.1990. *Domination and the art of resistance: hidden transcripts.* New Haven Conn.: Yale University Press.

Seers, Dudley. 1969. The meaning of development. Paper presented at the Eleventh World Conference of the Society for International Development, New Delhi, November, 4–17, 1969. *International Development Review*, 2, no. 6, 2–6.

Sen, Amartya. 1990. Gender and cooperative conflicts. In Irene Tinker, ed., *Persistent inequalities: women and world development*, chapter 8. New York: Oxford University Press.

Shanin, Theodor. ed. 1987. *Peasants and Peasant Societies*. 2d ed. London: Basil Blackwell.

Sharma, M. L. 1987. *Gandhi and democratic decentralization in India*. Delhi: Deep & Deep.

Sivan, Emmanuel. 1985. *Radical Islam: medieval theology and modern politics*. New Haven Conn.: Yale Univeristy Press.

Slater, David. 1989. Territorial power and the peripheral state. The issue of decentralization. *Development and Change*, 20, no. 3, 501–32.

Smith, Joan, et al., 1984. *Households and the world economy*. Beverly Hills: Sage Publications.

Soja, Edward. 1989. *Postmodern geographies: the reassertion of space in critical social theory*. London: Verso.

Soper, Kate. 1981. *On human needs: open and closed theories in a Marxist perspective*. Brighton, Sussex: Harverster Press.

Stanfield, J. R. 1986. *The economic thought of Karl Polanyi: lives and livelihood*. London: Macmillan.

Starr, John Bryan. 1979. *Continuing the revolution: the political thoughts of Mao*. Princeton, N.J.: Princeton University Press.

Stepan, Alfred. 1985. State power and the strength of civil society in the southern cone of Latin America. In Peter B. Evans, Dietrich Rueschmeyer, and Theda Skocpol, eds., *Bringing the state back in*. New York: Cambridge University Press.

Stöhr, Walter B., ed. 1990. *Global challenge and local response: initiatives for economic regeneration in contemporary Europe*. London: Mansell.

Stokes, Susan C. 1988. Peru's urban popular sectors in the 1980s: autonomy or a new multi-classism? Paper presented at the sixteenth Congress of the Latin American Studies Association, New Orleans, Louisiana.

Storper, Michael, and Walker, Richard. 1989. *The capitalist imperative. Territory, technology, and industrial growth*. New York: Basil Blackwell.

Streeten, Paul, and Burki, Shavid Javed. 1978. Basic needs: some issues. *World Development*, 6, no. 3, 411–21.

SUR. 1989. La reconstrucción democrática de Chile: la tarea de las Juntas de Vecinos y organizaciones comunitarias. *Hechos urbanos*, no. 90.

Sutton, J. Francis X. 1988. Development ideology: its emergence and decline. *Daedalus*, 118, no. 1, 35–67.

Tendler, Judith. 1983. *What to think about cooperatives: a guide from Bolivia*. In collaboration with Kevin Healy and Carol Michaels O'Laughlin. Rosslyn, Va.: Interamerican Foundation.

Tironi, Eugenio. 1987. Pobladores e integración social. *Proposiciones*, no. 14. *Marginalidad, movimientos sociales y democrácia*, pp. 64–84. Santiago: SUR.

Touraine, Alain. 1977. *The self-production of society*. Chicago: University of Chicago Press.

——. 1981. *The voice and the eye: an analysis of social movements*. New York: Cambridge University Press.

Tovar, Teresa. 1986. Barrios, ciudad, democrácia y política. In Eduardo Ballón, ed., *Movimientos sociales y democrácia: la fundación de un nuevo órden*. Lima: DESCO.

UNESCO. 1984. *Applicability of indicators of socio-economic change for development planning*. Paris: UNESCO.

United Nations. 1951. *Measures for the economic development of underdeveloped areas*. Report by a group of experts. New York: United Nations Organization, Department of Economic Affairs.

——. 1954. *Report on International definition and measurement of standards and levels of living*. Report of a committee of experts. New York: United Nations.

United Nations Development Program. 1990. *Human development report 1990*. New York: Oxford University Press.

Vanier Institute of the Family. 1983. *A social framework for economics: development from the ground up*. Submission of the Vanier Institute of the Family to the Royal Commission on the Economic Union and Development Prospects for Canada, Ottawa. Manuscript.

Veliz, Claudio. 1980. *The centralist tradition of Latin America*. Princeton N.J.: Princeton University Press.

Walton, John. 1989. Debt, protest, and the state in Latin America. In Susan Eckstein, ed., *Power and popular protest: Latin American social movements*, chapter 10. Berkeley: University of California Press.

Walzer, Michael. 1983. *Spheres of justice: a defense of pluralism and equality*. New York: Basic Books.

Weaver, Clyde. 1984. *Anarchy, planning, and regional development*. London: John Wiley & Sons.

Wilk, Richard K. 1989. *The household economy: reconsidering the domestic mode of production*. Boulder, Colo.: Westview Press.

Wolf, Eric R. 1982. *Europe and the people without history*. Berkeley: University of California Press.

Worby, Eric 1988. Livestock policy and development ideology in Botswana. In Donald W. Attwood, Thomas C. Bruneau, and John G. Galatay, eds., *Power and poverty: development and development projects in the Third World*. Boulder, Colo., and London: Westview Press.

World Bank. 1989a. *World development report 1989*. New York: Oxford University Press.

——. 1989b. *Striking a balance: the environmental challenge of development*. Washington, D.C.: IBRD.

——. 1990. *World development report 1990*. New York: Oxford University Press.

World Commission on Environment and Development. 1987. *Our common future*. Oxford: Oxford University Press.

World Resources Institute. 1986. *World resources 1986*. A report by the World

Resources Institute and the International Institute for Environment and Development. New York: Basic Books.

Young, Kate M. 1980. A methodological approach to analysing the effects of capitalist agriculture on women's roles and their position within the community. In International Labour Office, *Women in rural development: critical issues*. Geneva: ILO.

——. 1988. Introduction: reflections on meeting women's needs. In Kate Young, ed., *Women and economic development: local, regional and national strategies*. New York: Berg Publishers and Paris: UNESCO.

Index